About Island Press

Island Press is the only nonprofit organization in the United States whose principal purpose is the publication of books on environmental issues and natural resource management. We provide solutions-oriented information to professionals, public officials, business and community leaders, and concerned citizens who are shaping responses to environmental problems.

Since 1984, Island Press has been the leading provider of timely and practical books that take a multidisciplinary approach to critical environmental concerns. Our growing list of titles reflects our commitment to bringing the best of an expanding body of literature to the environmental community throughout North America and the world.

Support for Island Press is provided by the Agua Fund, The Geraldine R. Dodge Foundation, Doris Duke Charitable Foundation, The Ford Foundation, The William and Flora Hewlett Foundation, The Joyce Foundation, Kendeda Sustainability Fund of the Tides Foundation, The Forrest & Frances Lattner Foundation, The Henry Luce Foundation, The John D. and Catherine T. MacArthur Foundation, The Marisla Foundation, The Andrew W. Mellon Foundation, Gordon and Betty Moore Foundation, The Curtis and Edith Munson Foundation, Oak Foundation, The Overbrook Foundation, The David and Lucile Packard Foundation, Wallace Global Fund, The Winslow Foundation, and other generous donors.

The opinions expressed in this book are those of the author(s) and do not necessarily reflect the views of these foundations.

Markets and the Environment

Foundations of Contemporary Environmental Studies

James Gustave Speth, editor

Global Environmental Governance
James Gustave Speth and Peter M. Haas

Ecology and Ecosystem Conservation
Oswald J. Schmitz

Markets and the Environment
Nathaniel O. Keohane and Sheila M. Olmstead

Forthcoming:

Nature and Human Nature
Stephen R. Kellert

Environmental Law and Policy
Daniel Esty and Douglas Kysar

Human Health and the Environment
John Wargo

MARKETS
AND THE
ENVIRONMENT

Nathaniel O. Keohane
Sheila M. Olmstead

ISLANDPRESS

WASHINGTON · COVELO · LONDON

For Frances and Eleanor, and for Gau.
—N.O.K.

For my father, Joe Cavanagh, and my sons, Kevin and Finn.
—S.M.O.

Library of Congress Cataloging-in-Publication Data

Keohane, Nathaniel O.
   Markets and the environment / Nathaniel O. Keohane and Sheila M. Olmstead.
     p.   cm.
   Includes bibliographical references and index.
   ISBN-13: 978-1-59726-047-3 (pbk. : alk. paper)
   ISBN-10: 1-59726-047-9 (pbk. : alk. paper)
   ISBN-13: 978-1-59726-046-6 (cloth : alk. paper)
   ISBN-10: 1-59726-046-0 (cloth : alk. paper)
   1. Environmental economics.   I. Olmstead, Sheila M. II. Title.
   HC79.E5K422 2007
   333—dc22
                           2007014814

Table of Contents

Preface

This book provides a concise introduction to the economic theory of environmental policy and natural resource management. Like the other volumes in the series Foundations of Contemporary Environmental Studies, ours is a self-contained treatment of one aspect of the interaction between human society and the natural world. We have written the book with university students in mind, but the importance of the subject and the informal style of the book make it suitable for a wide range of professionals or other concerned readers seeking an introduction to environmental economics.

We wrote this book in the hopes that it will help to illuminate the role that economic theory—and more broadly economic *thinking*—can play in informing and improving environmental policy. To our minds, noneconomists tend to perceive economics rather narrowly—as being concerned only with money, or with national indicators like exchange rates and trade balances. In fact, economics has a much wider reach. It sheds light on individuals' consumption choices in the face of scarce resources; the interaction between firms and consumers in a market; the extent to which individuals are likely to contribute toward the common good, or ignore it in the pursuit of their own self-interest; and the ways that government policies and other institutions shape incentives for action. As we explain in the first chapter of the book, economics is central to understanding why environmental problems arise and how and why to address them. As concerned citizens as well as economists, we think that it is vital for anyone interested in environmental policy to be conversant in the language of economics.

The approach we have taken here draws upon our experience teaching environmental and natural resource economics to masters students and undergraduates. The emphasis is on intuition rather than algebra; we seek to convey the underlying concepts through words and graphs, instead of presenting mathematical results. We have also included a wealth of real-world examples—from the conservation of the California condor, to the reduction of CO_2 concentrations in the atmosphere, to the market-based model of fisheries management underway in New Zealand. The book is designed to be accessible to someone without any prior knowledge of economics. At the same time, the treatment is comprehensive enough that even an economics major could learn a great deal from the book. The lack of mathematical notation does not reduce the rigor of the underlying analysis.

In our teaching, we have noticed a yawning gap between short articles on "how economists think about the environment" and textbooks filled with algebra and detailed information on the history of U.S. federal environmental legislation. In addition, most textbooks on the subject of markets and the environment treat either the economics of pollution control or the economics of natural resource management. At an introductory level there is little integration of these two "halves" of the discipline of environmental and resource economics. This book aims to fill these gaps. We envision it as a primer to be used as part of a core course in environmental studies, either at the undergraduate or masters level. In that context, this book would be the sole economics text, used alongside several other books (ideally from the same series) representing different perspectives on environmental studies from the social and natural sciences. The book is also well-suited to a semester-long course in environmental or natural resource economics—either as a main text (supplemented with more mathematical lecture notes) or as a complement to another, more detailed (but perhaps less intuitive) textbook. Finally, the book could be used (as we ourselves have used the notes from which it grew) as an introduction to environmental economics in a course with a different focus. For example, a course on business strategy can use this book to explain the basic logic and practice of market-based policies to regulate pollution. Similarly, a Principles of Microeconomics course could use this book to show how economic theory can be applied to real-world problems and illuminate the market failures aspect of the course.

At the end of the volume, readers will find a list of references, including works cited in the text as well as other recommended readings of possible interest. We have also provided a set of study questions for each chapter, designed to be thought-provoking and open-ended rather than simply reiterating the material.

We thank Karen Fisher-Vanden for providing thoughtful comments on a draft. Our editor at Island Press, Todd Baldwin, helped us to find our voice and patiently moved us through the process of writing the book. Kartik Ramachandran's careful reading of the draft manuscript improved it immensely. Matt Finkelstein and Zack Donohew provided excellent research assistance. We thank our spouses, Todd Olmstead and Georgia Levenson Keohane, for their support and encouragement. Finally, we both owe a great deal to Robert Stavins, whose passion for teaching environmental economics—and unrivalled ability at doing so—continues to inspire us.

NOK
SMO
New Haven, Connecticut

1

Introduction

This book is a primer on the economics of the environment and natural resources. The title, *Markets and the Environment*, suggests one of our central themes. An understanding of markets—why they work, when they fail, and what lessons they offer for the design of environmental policies and the management of natural resources—is central to an understanding of environmental issues. But even before we start thinking about how markets work, it is useful to begin with a more basic question: What *is* environmental economics?

Economics and the Environment

"Environmental economics" may seem like a contradiction in terms. Some people think that economics is "just about money," that it is preoccupied with profits and economic growth and has nothing to do with the effects of human activity on the planet. Others view environmentalists as being naïve about economic realities, or "more concerned about animals than jobs."

Of course neither stereotype is true. Indeed, not only is "the environment" not separate from "the economy," but environmental problems cannot be fully understood without understanding basic economic concepts. Economics helps explain why firms and individuals make the decisions they do—why coal (dirty, polluting, high-in-carbon coal) is still the dominant fuel for electric power plants in the United States, or why individuals drive Hummers instead of Priuses. Economics also helps predict how those same firms and individuals will respond to a new set of incentives—for example, what investments electric utilities will make in a carbon-constrained world, and how high gas prices would have to rise before people stopped buying sport utility vehicles.

At its core, economics is the study of the allocation of scarce resources. This central focus, as much as anything else, makes it eminently suited to analyzing environmental problems. Let's take a concrete example. The Columbia and Snake rivers drain much of the U.S. Pacific Northwest, providing water for drinking, irrigation, transportation, and electricity generation—as well as for the support of endangered salmon populations. All of these activities—including salmon preservation—provide *economic* benefits to the extent that people value them.

If there is not enough water to meet all those needs, then we must trade off one good thing for another: less irrigation for more fish habitat, for example. How should we as a society balance these competing claims against each other? To what lengths should we go to protect the salmon? What other valued uses should we give up? We might reduce withdrawals of water for agricultural irrigation, remove one or more hydroelectric dams, or implement water conservation programs in urban areas. How do we assess these various options?

Economics provides a framework for answering these questions. The basic approach is simple enough: Measure the costs and benefits of each possible policy, including a policy of doing nothing at all, and then choose the policy that generates the maximum net benefit to society as a whole (that is, benefits minus costs). This is easier to say than to carry out, but economics also provides tools for measuring costs and benefits. Finally, economic theory suggests how to design policies that harness market forces to work *for* rather than against environmental protection.

To illustrate how economic reasoning can help us understand and address environmental problems, let's take a look at perhaps the most pressing environmental issue today, and one that is increasingly in the news: global climate change.

Global Climate Change

There is overwhelming scientific consensus that human activity—primarily the burning of fossil fuels and deforestation due to agriculture and urbanization—is responsible for a sharp and continuing rise in the concentration of carbon dioxide (CO_2) and other heat-trapping gases in the earth's atmosphere. The most direct consequence is a rise in average global surface temperatures, which is why the phenomenon is known widely as "global warming." (Surface temperatures have already increased worldwide by 0.6 degrees Celsius, or about 1 degree Fahrenheit, since the start of the twentieth century.)[1] But the consequences are much broader than warming, which is why the broader term "climate change" is more apt. Expected impacts (many of which are already measurable) include sea level rise from the melting of polar ice caps; regional changes in precipitation; the disappearance of glaciers from high mountain ranges; the deterioration of coastal reefs; increased frequency of extreme weather events like droughts, floods, and major storms; species migration and extinction; and spatial shifts in the prevalence of disease. The worst-case scenarios include a reversal of the North Atlantic thermohaline circulation—better known as the Gulf Stream—which brings warm water northward from the tropics and makes England and the rest of northern Europe habitable. While there has been much international discussion regarding the potential costs and benefits of taking steps to slow or reverse this process, little progress has been achieved.

What are the causes of climate change? A natural scientist might point to the complex dynamics of the earth's atmosphere—how CO_2 accumulating in the atmosphere traps heat (the famous "greenhouse effect"), or how CO_2 gets absorbed by ocean and forest "sinks." From an *economic* point of view, the roots lie in the incentives facing individuals, firms, and governments. Each time we drive a car, turn on a light, or use a computer, we are indirectly increasing carbon emissions and thereby contributing to global climate change. In doing so, we impose a small cost on the earth's population. These costs, however, are invisible to the individuals responsible. You do not pay for the carbon you emit. Nor, indeed, does the company that provides your electricity (at least if you live in the United States), or the company that made your car. The result is that we all put CO_2 into the atmosphere, because we have no reason not to. It costs us nothing, and we receive significant individual benefits from the energy services that generate carbon emissions.

Economics stresses the importance of incentives in shaping peoples' behavior. Without incentives to pay for the true costs of their actions, few people (or firms) will voluntarily do so. You might think at first that this is because the "free market" has prevailed. In fact, that gets it almost exactly backward. Very often, as we shall see in this book, the problem is not that markets are so pervasive, but that they are not pervasive *enough*—that is, they are incomplete. There is simply no market for clean air or a stable global climate. If there were, then firms and individuals who contributed to climate stabilization (by reducing their own carbon emissions or offsetting them) would be rewarded for doing so—just as firms that produce automobiles earn revenue from selling cars. This is a key insight from economics: Many environmental problems would be alleviated if proper markets existed. Since those markets don't arise by themselves (for reasons we shall discuss later on in the book), governments have a crucial role in setting them up—or in creating price signals that mimic the incentives a market would provide.

If this is such a problem, you may have asked yourself, why haven't the world's countries come together and designed a policy to solve it? After all, the consequences of significant climate change may be dire, especially for low-lying coastal areas and countries in which predicted changes in temperature and precipitation will marginalize much existing agricultural land. If you have been following the development of this issue in the global media, and you know of the difficulty experienced by the international community in coming to agreement over the appropriate measures to take in combating climate change, it will not be terribly surprising that economics predicts that this is a difficult problem to solve. Carbon emissions abatement is what economists would call a global *public good*: everyone benefits from its provision, whether they have contributed or not. If a coalition of countries bands together to achieve a carbon emissions abatement goal, all countries (including nonmembers of the coalition) will benefit from their efforts. So how can countries be induced to pay for it, if they will receive the benefit either way? This is a thorny problem to which we will return in later chapters.

As a starting point, we must understand just what the benefits of carbon emissions abatement are. They may be obvious to you. Put simply, slowing climate change can help us avert damages. For example, rising seas may inundate many coastal areas. If it is possible to slow or reverse this process, we might avoid damages including the depletion of coastal wetlands, the destruction of cultural artifacts, and the displacement of human populations. Warming in Arctic regions may lead to the extinction of the polar bear and other species; the benefits from slowing or reversing climate change would include the prevention of this loss. Climate change may exacerbate local pollution (such as ground-level ozone) and boost the spread of disease (like malaria in the tropics and West Nile virus in North America); we would want to measure the benefits from avoiding those damages, as well.

All of these benefits (even the intangible ones like species preservation) have economic value. In economic terms, their value corresponds to what people would be willing to pay to secure them. Measuring this value is relatively easy when the losses are reflected in market prices, like damages to commercial property or changes in agricultural production. But economists also have developed ways to measure the benefits of natural resources and environmental amenities that are not traded in markets, such as the improvements in human health and quality of life from cleaner air, the ecosystem services provided by wetlands, or the existence value of wilderness.

The economic cost of combating global climate change, meanwhile, is the sum of what must be sacrificed to achieve these benefits. Economic costs include not just out-of-pocket costs, but also (and more importantly) the foregone benefits from using resources to slow or reverse climate change, rather than for other objectives. Costs are incurred by burning cleaner but more expensive fuels, or by investing in pollution abatement equipment; by altering individual behavior, say by turning down the heat or air conditioning; by sequestering carbon in forests, oceans, depleted oil reservoirs, and other sinks; and by adapting to changing climatic conditions, for example by switching crops or constructing seawalls. Costs arise from directing government funds for research and development into climate-related projects rather than other pursuits. And of course the implementation, administration, monitoring, and enforcement of climate policy is costly in its own right.

Sound public policy decisions require an awareness of these costs and benefits, and some ability to compare them in a coherent and consistent fashion. Economics provides a framework for doing so. In practice, as you will see through the theory and examples in this book, implementing the framework requires taking account of a number of other wrinkles. For example, we must worry about how to weigh near-term costs against benefits that accrue much later.

There is one more question about climate change that economics can help answer: How should society address climate change? Even if the economist's prescription to maximize net benefits is ignored, economic reasoning can help improve policy design.

For example, suppose that the United States eventually signs on to a global climate change agreement, and commits to meet a particular abatement target (such as a certain number of tons of CO_2 reduced each year below some baseline). This target need not be efficient from an economic point of view;

it might well be shaped instead by purely political concerns. Regardless of how it is set, such a policy objective could be achieved in myriad ways—by requiring polluters to install and operate specific abatement technologies; by mandating tough energy efficiency standards for consumer appliances (and tightening fuel efficiency requirements for cars); by levying taxes based on CO_2 emissions; or by capping carbon emissions in an industry sector and allowing firms to trade allowances among themselves under that cap. (And that is hardly an exhaustive list!) As we will discuss at length in this book, especially in chapters 8 through 10, both economic theory and experience provide compelling arguments for "market-based policies," such as emissions taxes and cap-and-trade policies, that harness market forces to achieve regulatory goals at less overall cost than traditional approaches.

In sum, economics offers quite a different approach than other disciplines to the problem of global climate change—as well to as a range of other environmental issues we will explore in this book. You will find that the economic approach sometimes arrives at answers that are compatible with other approaches, and sometimes at answers that conflict with those approaches. Regardless of such agreement or disagreement, economics provides a set of tools and a way of thinking that anyone with a serious interest in understanding and addressing environmental problems should be familiar with.

Organization and Content of This Book

This book provides an introduction to the application of economic reasoning to environmental issues and policies. In each chapter, we draw heavily on a range of real-world examples to illustrate our points.

Chapter 2 begins by asking: "Why compare benefits and costs?" Here we introduce the central concept of economic efficiency, meaning the maximization of the net benefits of a policy to society. We illustrate the key points by discussing the abatement of sulfur dioxide at U.S. power plants, as well as a range of other examples. We introduce the key concepts of "marginal" costs and benefits, showing how they relate to total costs and benefits and how they inform the analysis of efficiency. We also extend the concept of efficiency to the dynamic context, in which policies are defined by streams of benefits and costs occurring over time. In doing so, we introduce the concept of discounting, the process by which economists convert values in the future to values today, and explain its usefulness in a dynamic setting.

Chapter 3 follows up on the same themes. We first discuss how economists define and measure the costs and benefits of environmental protection.

We then consider benefit-cost analysis in practice, discussing how it has been employed to evaluate policies in the real world. Finally, we explore the philosophical justification for benefit-cost analysis and consider some of the most frequent criticisms lodged against its use. In particular, benefit-cost analysis focuses on the net benefits from a policy, rather than its distributional consequences. Partly for this reason, economists do not advocate using a simple "cost-benefit test" as the sole criterion for policy decisions. While it is a valuable source of information, benefit-cost analysis is just one of a number of tools to use in assessing policies or setting goals.

We then turn our attention more explicitly to markets—how they function, what they do well and what they do poorly, and how they can be designed to achieve desirable outcomes. We begin chapter 4 with a key insight from economics: Under certain conditions, competitive markets achieve efficient outcomes. That is, they maximize the net benefits to society from the production and allocation of goods and services. This is a powerful result, and helps explain the wide appeal of markets. It also aids understanding of the root causes of environmental problems: to an economist, they stem from well-defined failures in how unregulated markets incorporate environmental amenities. Moreover, it lays the groundwork for designing policies that rely on market principles to promote environmental protection.

The notion of "market failure" is the focus of chapter 5. We discuss three ways of framing the types of market failure most common in the environmental realm: externalities, public goods, and the "tragedy of the commons". In each case we offer a range of motivating examples. We then unify the discussion by showing how each of the three descriptions of market failure captures the same underlying divergence between individual self-interest and the common good.

In chapter 6, we apply the concept of dynamic efficiency to the problem of the optimal rate of extraction of a nonrenewable natural resource, like petroleum. We define scarcity in economic terms, which leads naturally to the concept of rent—the extra economic value imparted by scarcity. We illustrate the underlying similarities between nonrenewable resources and other capital assets, and emphasize the powerful market incentives that encourage private owners of nonrenewable resources to account for scarcity in their extraction decisions.

Chapter 7 applies this same reasoning to two renewable resources, forests and fish. We develop bioeconomic models to demonstrate the efficient level of fishing effort and the efficient rotation period for a forest stand, both graphically and conceptually. In both cases, we include noncommercial benefits in an economic approach to efficient use of the resource.

Chapter 8 discusses the design of policies to overcome market failures in the provision of environmental amenities and the management of natural resources. We start by considering a central debate in economics, namely: Should the government intervene to solve market failures? After satisfying ourselves that the answer is yes, at least in many cases of real-world concern, we go on to review the various tools a government regulator has at her disposal, ranging from conventional "command-and-control" policies such as technology standards to market-based instruments such as taxes on pollution or resource use and tradeable allowances. We discuss the intuition behind how these latter approaches can restore the efficient workings of the market. We close by contrasting the two market-based instruments, asking when "prices" or "quantities" are the preferable tool for governments to wield.

Chapter 9 continues our discussion of policy design, but focuses more broadly on cases where efficiency may not be the objective. Even so, market-based instruments have two strong advantages: they can (in theory) achieve a desired level of environmental protection at the lowest total cost, while spurring the development and diffusion of new technologies over the long run. We briefly consider a range of other factors relevant to the design of policy. Market-based instruments are not the solution to every problem, and we show when conventional "command-and-control" approaches are preferable even on strictly economic grounds. But the main conclusion is that market-based instruments are a crucial component of the regulatory toolkit.

Chapter 10 reviews the real-world performance of market-based instruments in regulating pollution and managing natural resources. We consider three cases in careful detail: the market for sulfur dioxide emissions from power plants in the United States, the individual tradable quota system for New Zealand's fisheries, and municipal drought pricing of water resources in the United States. In each of these cases, we discuss the performance of the market-based approach, consider the implications for distributional equity, and assess the ease of monitoring and enforcement. We go on to review a longer catalog of examples, each in less detail than the initial case studies. Our aim is to equip readers to think broadly and creatively about the ways in which prices and markets can be injected into the regulatory process, aligning the incentives of firms and consumers with those of society in achieving environmental and resource management goals.

Chapter 11 addresses the links between economic growth and the natural environment—topics grouped under the heading of "macroeconomics," in contrast to the *micro*economic reasoning (based on the behavior of individuals and firms) that characterizes most of the book. We begin by reviewing the debate over the limits imposed on economic growth by natural resource scarcity, focusing on the critical importance of two often overlooked factors: substitutability and technological change. The same key issues arise in our discussion of "sustainability" in economic terms. We highlight the insights of economic definitions of sustainability for current natural resource management and environmental protection. We end with a discussion of "green accounting," emphasizing the need to incorporate natural resource depletion and changes in environmental quality into traditional measures of economic growth.

In the concluding chapter, we reflect on the relative roles of firms, consumers, and governments in the creation and the mitigation of environmental and resource management problems. We then offer some final thoughts about the role of economic analysis as one of many important tools at the disposal of decision-makers in environmental policy.

What We Hope Readers Will Take Away from This Book

If this is your first and last exposure to economics, and your interests lie in other areas of environmental studies, we offer three good reasons to use this text. First, many of the causes and consequences of environmental degradation and poor natural resource management are economic. That is, they arise from the failure of an unregulated market to give firms and individuals adequate incentives to promote environmental quality. Second, so-called market-based approaches to environmental regulation and natural resources management are increasingly common at local, national, and global levels. Prominent examples include the cap-and-trade policies already in place to limit sulfur dioxide pollution and greenhouse gas emissions, and tradeable fishing quotas to manage commercial fisheries. Third, economic arguments play an important role in some environmental policy debates, like management of public lands and the structure of international approaches to counter global climate change. Without an understanding of basic economic principles, it is difficult to formulate an economic argument—or to refute one.

Thinking systematically about benefits, costs, and tradeoffs can improve your ability to tackle real-world environmental problems—even when it is not possible to estimate benefits and costs explicitly. The theory we introduce and the applications we discuss are meant to demonstrate this. Of course, our treatment of individual topics in this text is necessarily brief; our intention is to give you just a basic grounding in the field. But we hope that the information we do present will pique your interest and prompt you to explore environmental and resource economics in greater depth.

This book alone will not make you an economist. Nonetheless, we hope to show that despite its reputation as a "dismal science," economics can make vital contributions to the analysis of environmental problems and the design of possible solutions.

2

Economic Efficiency and Environmental Protection

Imagine that you are planning a spring break trip to the Bahamas, and you are choosing from among four vacation packages you have found on the Web. The "Bahamas on a budget" trip, a three-day affair staying in tent cabins, costs $400. Suppose you would be *willing* to pay up to $550 for that trip, but no more. In other words, you wouldn't care if you paid $550 for the trip, or spent the money on something else. The next step up is a trip that costs $600. This trip includes four days' lodging in beachfront cabanas, and the setting is so beautiful that you would be willing to pay up to $900 for it. An even pricier five-day trip, with a few extras thrown in, would cost $850 and be worth $1100 to you. Finally, a deluxe week-long package is available for $1250, which on your student's budget is just about the maximum you would be willing to pay for any vacation, though this package is so breathtaking, you might just be willing to pay that much for it.

Faced with these possibilities, which trip should you choose? At first glance, you might think that the deluxe trip is the best one to take—after all, you value it the most, and are willing to pay the cost (even if only just barely). But in that scenario, you end up with zero *net* benefits. Indeed, since we have defined your "willingness-to-pay" as the amount for which you would be indifferent between paying for the trip and staying home, going on (and paying for) the week-long trip would make you no better off than if you didn't take a vacation at all! Choosing the deluxe trip on the grounds that you would be willing to pay the most for it amounts to ignoring the costs of the vacation completely.

Instead of choosing the trip with the highest *gross* value to you, regardless of cost, you would be better off choosing the trip that gives you the greatest *net* benefit—that is, the difference between the benefit of taking the trip (measured by your willingness-to-pay) and the cost (measured by its price). On these grounds, the best option turns out to be the four-day $600 trip, which you value at $900, for a net benefit of $300. This is greater than the net benefit from the more expensive $850 trip: the added cost (+$250) outweighs the increase in value (+$200), so that net benefits decline to $250. The $600 trip is also better (from a net-benefit perspective) than the "budget" trip. While that trip is cheaper, it is also worth less to you—and the drop in value is greater than the cost savings.

So how does this resemble an environmental problem? Well, imagine that, instead of taking a trip to the Bahamas, you are evaluating the possibilities for reducing pollution in your community, and there are a number of different options and price tags. As in the case of the vacation, a reasonable criterion for making decisions is maximizing net benefits. The net benefits of controlling air pollution, for example, are the difference between the total benefits of cleaner air and the total costs of reducing emissions. Maximizing the net benefits of a policy corresponds to the notion of economic efficiency. And as we'll see in chapter 3, willingness-to-pay is indeed at the heart of how economists conceive of and measure the value of environmental protection and natural resources.

You may be surprised to learn that if we accept economic efficiency as a reasonable goal for society, then the optimal level of pollution will in general be greater than zero! The reason for this will become clear as we proceed, but it can be summed up in a nutshell as follows: while there would certainly be benefits from eliminating pollution completely, the costs would (in most cases) be much higher. We could get nearly the same benefit, at much lower cost, by tolerating some pollution.

Economic Efficiency

To an economist, answering the question "How much environmental protection should society choose?" is much like answering the question "Which vacation package is best?" in the simple example above (albeit on a much larger scale): it depends on comparing benefits and costs, and finding where their difference is greatest.

This comparison between benefits and costs leads to a central concept in economics: that of *economic efficiency*. To an economist, an efficient policy or outcome is one that achieves the greatest possible net benefits. You should note that "efficiency" has a precise meaning here, which differs somewhat from common usage. In other contexts, "efficiency" connotes a minimum of wasted effort or energy. For example, the "energy efficiency" of a home appliance refers to the amount of electricity the appliance uses, per unit of output—for example, the amount of electricity used by an air conditioner to cool a room of a certain size. The less energy an appliance uses to produce a given outcome, the more energy-efficient it is. Similarly, the "efficiency" of a generator in an electric power plant measures how much useful energy a turbine generates, relative to the energy content of the fuel burned to drive the turbine. In both of these examples, "efficiency" is a function only of inputs and processes. The goal (cooling a room of a given size, or generating a certain amount of electricity) is taken as given, and "efficiency" measures how little energy is used to achieve it. In other words, energy efficiency does not relate benefits and costs—the comparison at the heart of the concept of economic efficiency.

To illustrate this contrast, suppose you are choosing between a top-of-the-line air conditioner that costs $500, and a model that uses more electricity but costs only $150. The more expensive air conditioner is certainly more energy efficient. However, whether it is more efficient from an economic point of view—that is, whether the net benefits are greater—depends on how often you will use the air conditioner, how much more electricity the lower-end model uses, and the price of electricity.

To understand what economic efficiency means for environmental policy, let's start by considering a real-world environmental issue: sulfur dioxide (SO_2) emissions from fossil-fueled electric power plants. Burning oil or coal to generate electricity creates SO_2 as a by-product, because those fuels contain sulfur. In downwind areas, SO_2 emissions contribute to urban smog, particulate matter, and acid rain. For these reasons, the control of SO_2 emissions from power plants has been a focus of air pollution legislation in the United States and many other countries.

From an economic perspective, we can frame this issue in terms of the efficient level of SO_2 emissions abatement. (It is often easier to think in terms of abatement, or pollution control, which is a "good," rather than pollution, which is a "bad.") Suppose we observe the amount a firm or industry would pollute in the absence of any regulatory controls. Abatement is measured relative to that benchmark. If a firm would emit

To an economist, being efficient means maximizing net benefits.

a thousand tons of pollution in the absence of regulation, but cuts that to six hundred tons of pollution (for example, by installing pollution control equipment), it has achieved four hundred tons of abatement.

What level of sulfur dioxide abatement will maximize net benefits to society? To answer this question, of course, requires thinking systematically about the costs and benefits of pollution control.

The Costs of Sulfur Dioxide Abatement

Typically, a minor amount of abatement can be achieved at very little cost simply by improving how well a power plant burns coal, since a cleaner-burning plant will emit less pollution for any given amount of electricity generated. (One reason the resulting abatement is cheap is that a cleaner-burning plant will also use up less fuel to produce the same amount of electricity—saving money for its managers.) At a somewhat higher cost, power plants can increase their abatement by burning coal with slightly less sulfur than they would otherwise use. The abatement cost increases further as the power plant burns coal containing less and less sulfur that is more and more expensive. For example, a power plant in Illinois can burn relatively cheap high-sulfur coal from mines in the southern part of the state. To reduce SO_2 pollution, such a plant might switch to coal from eastern Kentucky with half the sulfur content but a slightly higher transportation cost. Still greater reductions could be achieved, at still greater cost, by switching to very low sulfur coal from Wyoming. Finally, to achieve reductions of 90 percent or more from baseline levels typically requires investment in large "end-of-pipe" pollution control equipment, such as flue-gas desulfurization devices (better known as "scrubbers") that remove SO_2 from the flue gases. Such equipment is often very expensive, making high levels of abatement much more costly than low levels. Moreover, the cost is typically driven by the percentage reduction achieved, so that removing the first 90 percent of pollution costs about the same as going from 90 to 99 percent removal.

The costs we just described trace out a particular pattern. Costs rise slowly at first, as abatement increases from zero. As abatement continues to increase, however, costs rise more and more rapidly. This pattern is reinforced when we consider the costs of abatement at the level of the industry, rather than the individual power plant. Some power plants (those located relatively close to low-sulfur coal deposits, for example) can abate large amounts of pollution at relatively low cost, while others may find even small reductions very expensive. As we increase pollution control at the industry level, we must call on plants where abatement is more and more expensive.

Figure 2.1 depicts a stylized *abatement cost function* that corresponds to this pattern of rising cost. By "abatement cost function," we mean the total cost of pollution control as a function of the amount of control achieved. On the figure, we have used X to represent the amount of pollution control, and $C(X)$ to denote the total cost (in dollars) as a function of X. A function with this bowed-in shape is called a convex function.

The Benefits of Sulfur Dioxide Abatement

Recall that in the first chapter we described the benefits from reducing greenhouse gas emissions as corresponding to the avoided damages from global climate change. In the same way, the benefits of SO_2 abatement are simply the avoided damages from pollution.

How do these damages vary with pollution? As the air gets dirtier, pollution damages tend to increase more and more rapidly. At low concentrations, SO_2 corrodes buildings and monuments. Higher concentrations lead to acid rain, with the attendant damages to forest ecosystems from the acid-

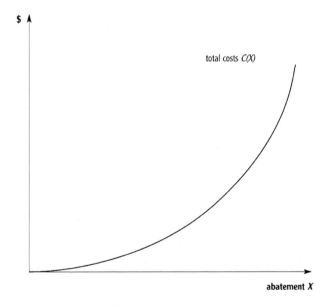

Figure 2.1 Total costs of pollution abatement, as a function of the level of abatement.

ification of lakes and soils. In urban areas, the adverse effects of SO_2 increase from eye and throat irritation, to difficulty breathing, and ultimately to heart and respiratory ailments. These effects are felt first by the most vulnerable members of society: infants, the elderly, and asthmatics. But as concentrations rise, the affected population grows.

This pattern of damages corresponds to total *benefits* from pollution control that increase rapidly when abatement is low (and pollution is high), and increase more slowly when abatement is high (and pollution is low). This is illustrated by the curve in figure 2.2, where we have used $B(X)$ to represent the *abatement benefit function*. A function with the bowed-out shape of $B(X)$ is called a concave function.

Putting Costs and Benefits Together: Economic Efficiency

We are now ready to answer the question we posed earlier: What is the efficient level of sulfur dioxide abatement? To answer this question, we must compare benefits to costs, and find where the difference between them—net benefits—is greatest.

Figure 2.3 places the cost and benefit curves drawn above on a single pair of axes. As in the previous figures, abatement increases as we move along the

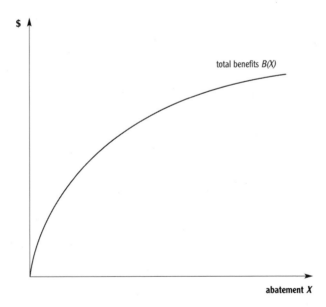

Figure 2.2   Total benefits of pollution abatement, as a function of the level of abatement.

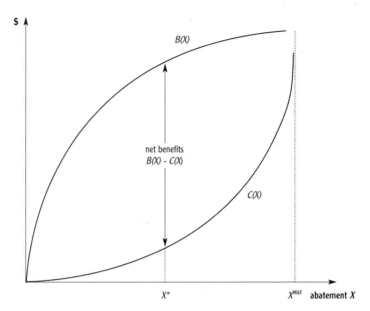

Figure 2.3   The efficient level of pollution abatement, denoted X^*, achieves the greatest possible net benefit.

horizontal axis from left to right; pollution increases as we move from right to left. We have denoted "maximum abatement"—equivalent to zero pollution—by X^{MAX}.

Recall that net benefits are simply benefits minus costs. Thus on the figure, the net benefit from a given level of pollution control is measured by the vertical distance from the benefit curve down to the cost curve. At low levels of pollution control, net benefits are small. As abatement increases from a low level, the benefits increase more rapidly at first than do the costs, so that net benefits increase. As more and more abatement is done, however, the benefits rise less rapidly, while the costs of abatement increase. Eventually, the benefits increase more slowly than costs, and net benefits *fall* as more and more abatement is done.

In between those two extremes, of course, the difference between benefits and costs must reach a maximum. On our graph, this happens at the level $X^\star$. By definition, this is the efficient level of pollution control. You can see from the figure that $X^\star$ is greater than zero, but less than the maximum possible abatement. Accordingly, the efficient level of pollution must also be less than its maximum (unregulated) level, but greater than zero. We come right away to the point that we mentioned at the outset of the chapter:

In general, the economically efficient level of pollution is not zero.

Zero pollution is not efficient (in general), because the gains from achieving it are not worth the extra cost required. Consider increasing abatement from the level $X^\star$ to the level X^{MAX}. In our real-world example, this might correspond to installing expensive scrubbers on every power plant. This much abatement would certainly bring benefits, such as reductions in acid rain and improvements in urban air quality. On the graph, the increase in benefits is shown by the fact that the curve $B(X)$ increases as we move to the right, so that $B(X^{MAX}) > B(X^\star)$.

However, those extra benefits from maximizing abatement are outweighed by the extra costs of achieving them. While benefits increase, costs rise even faster. As a result, the gap between benefits and costs shrinks dramatically as we increase abatement from $X^\star$ to X^{MAX}. In the real world, requiring scrubbers on all power plants would raise costs by an order of magnitude, while the boost in benefits would be much smaller.

Therefore, zero pollution is generally not desirable—at least not if we measure the success of our policy by the magnitude of its net benefits. Of course, it is equally true (although perhaps less surprising) that zero abatement is also not efficient. Abating less than $X^\star$ would reduce costs—but the cost savings would be less than the foregone benefits. On balance, net benefits would fall.

If you find these results surprising or counterintuitive, it may help to recall the distinction between economic and technical notions of efficiency. Pollution is sometimes described as "inefficient" when the pollution represents a form of wasted inputs. For example, a key component of water pollution from paper mills or textile factories is excess chemicals used in the production process—bleach in the case of paper mills, or dye in the case of textile factories. While such pollution may be "inefficient" in a technical sense, it is a mistake (albeit a common one) to conclude that it is also necessarily inefficient in an *economic* sense. Economic efficiency depends on the costs as well as the benefits of controlling pollution. If it is extremely costly to clean up pollution completely, zero pollution is unlikely to be a reasonable goal if we aim to maximize net benefits.

Efficiency and Environmental Policy

In our example of SO_2 pollution from power plants, the benefits from abatement rise rapidly at first and then tail off, while costs rise much more slowly at first before becoming steep. Put them together, and we find that net benefits are greatest somewhere in the middle. Because the shapes of the cost and benefit curves are critical in driving the results, it is worth discussing them in a bit more detail.

The pattern of "increasing costs at an increasing rate" is common. The cost of producing most goods—for example, steel or shoes—typically increases with production at an increasing rate (at least in the short run, and over some range of quantities). In the case of pollution control, you can think of "clean air" as the good that is being produced: clean air is costly, and the cost rises more and more steeply as the air gets cleaner and cleaner. Removing the last few ounces of pollution from a waste stream is likely to be prohibitively expensive.

On the benefit side, meanwhile, assuming a concave benefit function corresponds to the simple idea that while we would usually like more of a good thing, the amount we are willing to pay for something is likely to decline as we get more of it. You would probably pay more for one pair of designer shoes, or one pair of tickets to a rock concert, than you would pay for the second, third, or tenth pair of the same item.

These characteristics of costs and benefits apply in a wide range of cases in the environmental realm—not just other forms of air pollution, but also water pollution, biodiversity preservation, endangered species protection, the management of natural resources such as fisheries, and so on. For example, consider the protection of habitat for an endangered species such as the red-cockaded woodpecker, which lives in old-growth stands of longleaf pine forest in the southeastern United States. Habitat protection requires managing forests to maintain suitable old-growth conditions. The cost of such management varies widely among different parcels of land, depending on ownership, suitability for intensive timber production, soil conditions, and so on. If we arrange lands from least to greatest expense, we can construct an increasing cost-of-protection function similar to the one in figure 2.1. Similarly, on the benefits side, an increase in the woodpecker population from one hundred birds to two hundred birds is likely to yield much greater benefits than from one thousand birds to eleven hundred birds—leading to a benefit-of-protection function much like the curve in figure 2.2.

Accordingly, while we will continue to discuss our model in terms of pollution control or abatement, you should keep in mind that it is much more general than that. For convenience, we will continue to refer to X as pollution control or abatement; but you could substitute any other dimension of environmental quality, such as "habitat protection," and the arguments that follow would still apply. The crucial assumptions underlying our model are that costs increase at an increasing rate and that benefits increase at a decreasing rate—in other words, that the *total cost function* $C(X)$ is convex and the *total benefit function* $B(X)$ is concave, like those drawn in figures 2.1 through 2.3.

In some cases, of course, these assumptions do not hold. For example, think of litter along a hiking path in a wilderness area. One piece of trash may ruin an otherwise pristine area nearly as much as ten or twenty pieces would. In this case, the marginal benefit of environmental quality does not fall as the amount of trash gets smaller (until the trash goes away completely). Hence the efficient level of litter might well be zero.

A particularly important exception to the conventional rule "equate marginal benefit and cost" arises when the marginal cost of cleanup falls (instead of rising) as more cleanup is done. Cost functions with this characteristic are said to exhibit *economies of scale*. For example, cleaning up hazardous waste sites typically requires digging up the soil and incinerating it to remove the pollution. The cost of such a cleanup depends mostly on the area of the site, rather than how contaminated it is or how much pollution is removed. In such a case, pollution control may be an "all-or-nothing" exercise: if it makes sense to clean up a site at all, then it will make sense to clean it up completely. Over time, this policy would look very different than in the standard case of increasing marginal cost. Rather than seeking to maintain environmental quality at the level where marginal cost and benefit are equal, the optimal policy would let quality decline over time and then periodically clean things up to a very high level of quality.

Even if the cost and benefit functions have their typical shapes, of course, one can draw particular examples in which the maximum level of abatement is reached before net benefits start declining—or, conversely, in which net benefits are highest when abatement is zero. (Imagine taking the curves drawn in figure 2.3 and shifting them rightward or leftward while holding the axes and the location of maximum abatement fixed.) However, there are good reasons to view these instances as special cases, as we have already seen. The model of convex costs and concave benefits presented here is widely accepted as the conventional general model of the costs and benefits of pollution control (and of environmental protection more generally).

Equating Benefits and Costs on the Margin

So far, we have discussed the *total* costs and benefits of pollution control. An alternative and very useful way to describe the costs and benefits of pollution control is in terms of *marginal* costs and benefits. By marginal cost, we simply mean the cost of an incremental unit of abatement. If we have abated one hundred tons, the marginal cost is the cost of the one-hundredth ton. (Note the contrast with average cost, which takes into account all of the abatement done rather than only the last unit.) Likewise, marginal benefit refers to the benefit from the last unit of abatement. Recall that efficiency corresponds to maximizing the difference between total benefits and costs. It turns out that this difference is greatest when marginal benefit and marginal cost are equal.

Marginal Costs and Marginal Benefits

Let's start by considering the relationship between total cost and marginal cost. Since marginal cost measures the cost of one more unit of abatement, it corresponds to the *slope* of the total cost function. To see why this makes sense, consider the cost function depicted in figure 2.1. At low levels of abatement, where the total cost function is nearly flat, the height of the curve changes little as pollution control increases. Therefore, each additional unit of pollution control adds a relatively small amount to the total cost. In other words, the marginal cost of pollution control is small. At higher levels of abatement, the total cost function is steep, so that the cost rises rapidly as abatement increases. This means that the incremental cost of pollution control—the marginal cost—is high.[1]

Figure 2.4 plots the *marginal cost function* corresponding to a total cost function like that in figure 2.1. As before, abatement is on the horizontal axis; but now the vertical axis measures marginal rather than total cost. Thus the height of the curve $MC(X)$ at any given point represents the cost of each additional unit of abatement. Saying that abatement cost increases at an increasing rate is the same thing

A very useful way to describe the costs and benefits of pollution control is in terms of marginal— that is, incremental—costs and benefits.

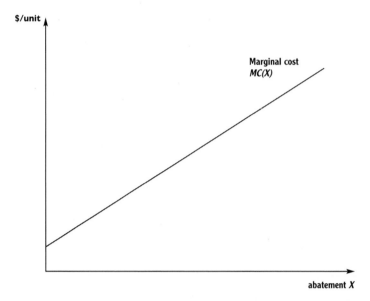

Figure 2.4 Representative marginal abatement cost function.

as saying that abatement has *increasing marginal costs*: each ton of pollution abatement costs slightly more than the one that preceded it. As a result, the marginal cost function in figure 2.4 slopes upward.

In a similar fashion, we can derive a *marginal benefit function* that corresponds to the incremental benefits of additional abatement. Marginal benefit corresponds to the slope of the total benefit function. If the benefit function is concave, as in figure 2.2, then the marginal benefit function will be downward sloping: each additional ton of abatement brings smaller additional benefits. We have drawn a representative function, labeled $MB(X)$, in figure 2.5.

Efficiency and the "Equimarginal Rule"

Let's take another look at figure 2.3, where we plotted the benefit and cost functions and found the efficient level of abatement X^*. Notice that as abatement increases up to X^*, the benefits of pollution control rise faster than the costs. That is, the $B(X)$ curve is steeper than the $C(X)$ curve. As a result, net benefits increase with each additional ton of pollution control over this range. On the other hand, beyond the efficient point, the costs rise faster than the benefits, so that net benefits diminish. Putting these obser-

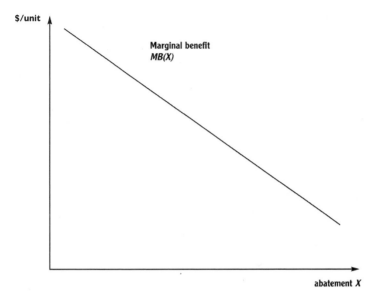

Figure 2.5 Representative marginal abatement benefit function.

vations together, we conclude that at the efficient level of abatement, the benefit and cost curves must have the same slope.

This suggests a way to find the efficient level of pollution control by looking at the marginal benefits and costs. In particular, we can state the *equimarginal rule*:

> The efficient level of abatement $X^\star$ occurs where marginal benefit equals marginal cost: that is, $MB(X^\star) = MC(X^\star)$.

In plain English, this says that the efficient level of pollution control is where the extra benefit of the last unit of abatement done equals its extra cost. Beyond that point, the additional costs of any further abatement will outweigh the benefits. This result is illustrated by figure 2.6. The top panel is the same as figure 2.3. The bottom panel draws the corresponding marginal benefit and cost curves. The efficient point $X^\star$ is easily identified: it is where the *MB* and *MC* curves cross.

This equimarginal condition will show up again and again in our analyses of markets and policy design. Thus it is worthwhile going over the intuition behind the result. Suppose we pick a low level of abatement, where

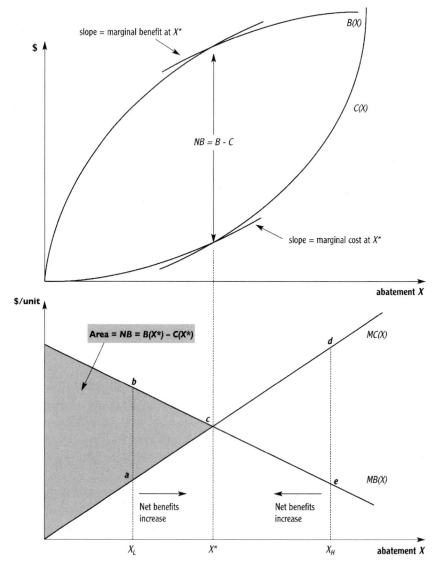

Figure 2.6 The efficient level of abatement, represented in terms of total costs and benefits (top panel) and marginal costs and benefits (bottom panel).

Thinking on the Margin: Pollution Abatement at Aracruz Celulose, S.A.*

One of the mainstays of economic reasoning is learning to think in terms of marginal changes when making decisions. To find the level of production that maximizes its profits, for example, a firm needs to compare the revenue from selling one more unit of the good with the cost of making it. Similarly, to find the amount of abatement that maximizes net social benefits, we must compare the marginal benefit from controlling another ton of pollution with the marginal cost.

To make the concept of marginal cost (in particular) more concrete, consider the case of pollution abatement at pulp mills owned by Aracruz Celulose, S.A., a leading Brazilian pulp producer and exporter. Among the major pollutants in effluent from pulp mills are chlorinated organic compounds, known as AOX (Adsorbable Organic Halides). These compounds—dioxin is among the most infamous—are produced when chlorine-containing chemicals used in bleaching react with wood fiber.

In the early 1990s, Aracruz was considering whether to upgrade its environmental controls in order to market its pulp to environmentally conscious customers in Europe. The table below lists four possible alternatives: doing nothing, switching to "elemental chlorine free" methods using chlorine dioxide, adding an oxygen delignification stage to reduce lignin content in the pulp before bleaching, and eliminating chlorine entirely ("totally chlorine free") by using peroxide as a bleaching agent. Note that these are cumulative efforts: for example, oxygen delignification is a necessary prerequisite to achieving TCF bleaching. In each case, we have given the resulting pollution level, the corresponding abatement, the total cost, and the marginal cost—that is, the cost per additional unit of abatement.

MB is greater than MC—say the point X_L on figure 2.6. Now let's imagine increasing abatement by one ton. What happens to net benefits? Since the resulting increase in benefits (equal to the marginal benefit) is greater than the increase in cost (= marginal cost), net benefits would increase. Thus at X_L efficiency increases with more abatement, as indicated by the arrow on the figure.

Now suppose that we increase abatement all the way to some high level, such as X_H on the figure where MB lies below MC. Here, one more ton of

Thinking on the Margin *continued*

Alternative	Pollution (AOX, in kg per year)	Incremental abate- ment (kg per year)	Total annual cost	Increase in total annual cost	Marginal cost (per kg of additional abatement)
1. Standard pulp (baseline)	1,000,000	No reduction	$0	$0	$0
2. ECF pulp using chlorine dioxide	250,000	750,000	$28.5 million	$28.5 million	$3.80
3. ECF + oxygen delignification	200,000	50,000	$29.6 million	$1.1 million	$22
4. TCF pulp using ozone	10,000	190,000	$40.4 million	$10.6 million	$56

* The cost figures for Aracruz are derived from estimates in Jackie Prince Roberts, "Aracruz Celulose, S.A.," Harvard Business School Case 9-794-049 (January 1995), 28 pp. In particular, the numbers in the table draw on the estimated costs for upgrading Mill A, Line 1. For purposes of illustration we have ignored the possible cost savings due to oxygen delignification and have treated that as an optional additional step. Abatement figures are based on 250,000 tons of pulp output per year and current authors' estimates of 4 kg AOX/ton for standard pulping, 1 kg/ton for ECF, 0.8 kg/ton for ECF + oxygen delignification, and 0.03 kg/ton for TCF. Following Roberts, we have applied a 10% discount rate to capital costs in order to combine them with variable costs.

abatement increases costs by more than it increases benefits: the incremental net benefit, therefore, is negative. Indeed, at such a point we could increase net benefits by reducing abatement by one unit, since costs would fall by MC, but benefits would decline by only MB. Thus at X_H we have overshot the efficient level of abatement.

Of course, we could repeat these arguments for any values of abatement above or below the point where the marginal curves cross. Only at the efficient point X^*, where $MB = MC$, is the difference between benefits and costs at its maximum.

Relating Marginal Benefits and Costs to Total Benefits and Costs

We have just seen how marginal benefits and costs correspond to the slopes of the total benefit and cost functions. Conversely, total benefits and costs can be represented as the *areas under the marginal benefit and cost curves*. Recall that the height of the marginal benefit curve (for example) at a given level of abatement represents the additional benefit derived from that unit of abatement. Imagine drawing a rectangle with width equal to one unit of abatement and height equal to the height of the *MB* curve. The area of that rectangle would be equal to the marginal benefit of the corresponding unit of abatement.

Now imagine drawing a series of such rectangles—one for each unit of abatement, starting from zero and going up to X_L. Since the area of each rectangle represents the additional benefit from a certain unit of abatement, their areas must sum to the total benefit from X_L units of abatement. But the sum of the areas of the rectangles is also equal to the area under the curve.[2] Thus the area under the marginal benefit curve from zero to any point equals the total benefit from that amount of abatement. Similarly, the area under the marginal cost curve from zero to any point is the corresponding total abatement cost.

This relationship between marginals and totals can give us another perspective on the "equimarginal condition" for efficiency. Let's return to the bottom panel of figure 2.6. At the efficient level of abatement (the point $X^\star$), total benefits equal the area under the *MB* curve, while total costs are the area under the *MC* curve. Subtracting costs from benefits leaves total net benefits (the shaded triangle to the left of the intersection of the two curves). You can see right away that no other level of abatement provides as much net benefit as $X^\star$. Less abatement leaves some net benefits unrealized. At X_L, for example, net benefits are smaller than at $X^\star$ by the area of the triangle labeled *abc* on the figure. Beyond $X^\star$, the extra costs outweigh the extra benefits. At X_H, net benefits are smaller than they are at $X^\star$ by the triangle *cde*.

Dynamic Efficiency and Environmental Policy

So far, we have discussed the efficiency rule of thumb—set marginal costs equal to marginal benefits—in terms of maximizing the net benefits of a resource (such as clean air or water) at a particular point in time. But projects and policies often have streams of benefits and costs occurring at many different points in time. For example, if we choose to set aside a large tract of land, like the Arctic National Wildlife Refuge in Alaska, disallowing commercial uses in favor of wilderness and recreation, society will receive benefits and incur costs from this designation over many years, or even in perpetuity.

When benefits and costs vary over time, economic analysis must apply the rules of *dynamic efficiency*.[3] For example, policies to address pollutants that accumulate in the environment—like carbon dioxide in the earth's atmosphere or polychlorinated biphenyls (PCBs) in a riverbed—involve streams of benefits and costs over a long period of time. Dynamic efficiency plays a particularly important role in the management of natural resources. Some resources, like petroleum, do not regenerate at all (at least over time scales relevant to human activity); for others (like fisheries), natural regeneration must be balanced against extraction and consumption. In both cases, the limited availability of the resources means that the amount available tomorrow depends on what we consume today. In order to apply the concept of efficiency in a dynamic setting, we must introduce the concept of *discounting*.

Discounting and Present Value

The introduction of a time dimension requires an additional step in thinking about efficiency. In the static analysis earlier in this chapter, we maximized net benefits. In a dynamic setting, an efficient policy maximizes the *present value* of net benefits to society. That is, we must convert all of the benefits and costs of a potential environmental policy, no matter when in time they occur, into their dollar value *today* before summing them up. In this way, we use a common yardstick to measure benefits and costs occurring at different points in time. To see why the value of one dollar today is not the same as the value of one dollar received next year, consider the following thought experiment. If we asked you whether you preferred a hundred dollars today or the same amount fifty years from now, you would certainly ask for the money today because you could invest it and earn a return over the next half century. For the same reason, if you owed us a hundred dol-

When benefits and costs vary over time, economic analysis must apply the rules of dynamic efficiency.

lars, you would rather pay us in fifty years than pay us today (and we would not agree to those terms!).

This important concept—the time value of money—is the reason that we "discount" costs and benefits expected to occur in the future. When we discount future benefits and costs, we ask whether the returns to a project, policy, or other investment, like a greenhouse gas emissions regulation, the establishment of a new national park, or the decision to pump groundwater from a nonrenewable aquifer, are greater or less than the returns to an alternative investment, such as placing the funds in an interest-bearing bank account, investing in education, or building a new hospital. If the answer to this question is "no," we can do better by choosing that alternative investment today, and letting future generations decide how to invest the returns.[4]

Discounting reduces a stream of costs and benefits to a single dollar amount—the present value (PV) of the stream—so that the PV of one policy can be compared to the PV of another. Discounting is the exact opposite of *compounding*, which converts PV to future value (FV). Suppose you invested a hundred dollars at an annual interest rate of 5 percent; how much would that investment be worth in fifty years? We can calculate the FV as follows, where r is the interest rate and t is the year

$$FV = PV(1+r)^t = 100(1+.05)^{50} = \$1,146.74.$$

(This is why we would not be willing to accept your offer to pay us a hundred dollars in fifty years, if you owed us that much today!)

You are likely familiar with the "power of compound interest." Discounting requires thinking in reverse. In the simple example above, if we wanted to know how much we needed to invest today to have $1,146.74 in the bank in fifty years, the answer, obtained by discounting, would be $100. We can also apply this kind of thinking to environmental problems. For example, the present value of a greenhouse gas emissions abatement policy that would cost one hundred dollars today, yielding annual benefits of fifty dollars per year over the next twenty years, would be calculated as follows:

$$PV = \sum_t \frac{FV_t}{(1+r)^t} = -100 + \frac{50}{(1+.05)^1} + \frac{50}{(1+.05)^2} + \dots + \frac{50}{(1+.05)^{20}} = \$812.80$$

For a hundred dollars invested today, we would receive today's equivalent of more than nine hundred dollars in benefits for a net present value of just over eight hundred dollars.

The Equimarginal Rule in a Dynamic Setting

The equimarginal rule that we discussed previously still applies in the dynamic setting, although we must convert marginal benefits and marginal costs into present value terms in order to compare the magnitude of streams of benefits and costs over time. In a dynamic context, efficient environmental policy equates the *present value of marginal benefits* with the *present value of marginal costs*. Typically, costs are borne today to generate benefits in the future, as in the hypothetical emissions abatement policy described above. In this case, the costs of the policy are not discounted (since they are already in present value), and the benefits are. We will explore other applications of the dynamic equimarginal rule in chapter 6, when we approach the problem of nonrenewable resource extraction, and in chapter 7, when we discuss the economics of forests and fisheries.

Conclusion

This chapter has laid the groundwork for everything that follows. When economists talk of efficiency, they have something very specific in mind: maximizing net benefits. As we have seen, in a static setting net benefits are largest (in general) when the benefits and costs of environmental protection are equal *on the margin*. In a dynamic setting, net benefits are largest when we equate marginal benefits and marginal costs in present value. This equimarginal condition represents a powerful tool for making decisions. In many instances, the benefits of taking some action (controlling pollution, say, or providing habitat for endangered species) are increasing at a decreasing rate, while the costs rise more and more rapidly. If so, the proper response—at least if we want to maximize net benefits—is to act until the benefit of one more unit of environmental quality just equals the incremental cost. In many cases, moreover, the benefits from pursuing "perfect" policies—such as zero pollution—often will not outweigh the costs. As a result, zero pollution is typically not an efficient outcome (although the same can also be said for zero pollution control).

The discussion in this chapter has abstracted from many of the challenges in using efficiency as a guide to policy. In particular, we have assumed that the costs and benefits of environmental protection are known, and we have taken for granted that maximizing net benefits is a reasonable goal to pursue. The next chapter tackles these challenges head-on.

3

The Benefits and Costs of Environmental Protection

The previous chapter proposed a destination—economic efficiency—for our journey, but it didn't give us a road map or even a compass. Imagine you are a policy maker deciding whether to approve construction of a hydroelectric dam on a wild river. Even if you embrace the idea of maximizing net benefits to society, how can you measure the costs and benefits of the project? How can you weigh a cheap, clean source of electricity against the damage to fish populations and the loss of rapids for rafting? How should you decide whether to build the dam or let the river run wild?

A first step is to define the costs and benefits of each option. To compare these costs and benefits, you need to measure them on a common yardstick. Then you can decide which option offers the greatest net benefit, though you may still want to ask why maximizing net benefits is what you should care about in the first place.

This chapter tackles these issues. We start by considering how economists think about the costs and benefits of environmental protection, and how those might be measured. While costs are relatively straightforward, benefits require considerable thought, as we shall see. We then consider how efficiency is implemented in practice, through "benefit-cost analysis" (the term is perhaps a bit more friendly than its common variant, "cost-benefit analysis"). That discussion culminates in an investigation into why efficiency might (or might not) be a desirable goal for policy.

Measuring Costs

How are costs defined and measured? In economic terms, the true costs of any activity are the *opportunity costs*—what you give up by doing one thing instead of another. For example, the true cost of going to graduate school is not simply the tuition plus the cost of room and board, but also (and crucially) the foregone income from two or more years out of work. More broadly, the prices of inputs such as capital, labor, and materials reflect their values in alternative uses. To produce electricity requires capital to pay for the construction of the generating unit (money that could have gone into alternative investments), labor to operate the plant (workers who could earn wages in other jobs), and fuel to produce steam (fuel that could have been used by other companies, and which required expending other resources in extraction and transportation). The same principle applies to reducing pollution: scrubbing sulfur dioxide out of flue gases requires capital to build the scrubber, and labor and materials to operate it. Devoting these resources to pollution control leaves less to spend on other opportunities, such as improving the plant's operation or increasing output.

Economics offers another valuable insight into the costs of environmental protection: they are ultimately borne by individuals, whether taxpayers, shareholders, or consumers. It is tempting to think that the benefits of clean air are enjoyed by society as a whole, while the costs of pollution control are paid out of corporate profits. In reality, of course, the costs of pollution control—even when they are "paid for" by corporations or electric utilities—end up being borne largely by consumers of the goods and services that cause the pollution. For example, electric utilities typically recover much of the cost of pollution control by charging higher rates for electricity. Even if abatement costs reduce the utilities' profits, much of that loss is felt by shareholders, who include retired pensioners as well as wealthy investors.

Two broad approaches are used to measure the costs of pollution control. The first relies on information that firms in many industries are required to report to the government. For example, electric power plants in the United States must fill out a Department of Energy survey that includes questions about annual expenditures on environmental protection. The second approach uses data on revenues and production costs to estimate the lost profits or diminished productivity associated with environmental protection—taking advantage of the fact that the profits foregone are equivalent to the costs incurred.

Outside of pollution control, the "costs" of environmental quality are often less obvious, but no less important to take into consideration. For example, let's consider endangered species protection. The costs of endangered species protection might include money spent on preserving habitat, enforcing prohibitions on hunting or poaching, and educating landowners and the public at large. Given information on how much these measures would cost and how effective they would be, we could estimate a mar-

In economic terms, the true costs of any activity are the opportunity costs—what you give up by doing one thing instead of another.

ginal cost function by dividing the cost of each action by the number of animals saved and arranging the actions in order of increasing cost.

Evaluating Benefits

Before we think about how to measure benefits, we must first think about how to define them. Let's start with an example. Suppose a parcel of open space near where you live—a wetland or a woodland or a beachfront site—is up for sale, and likely to be converted into a housing development. How would you think about your own value for preserving the open space? It might seem self-evident that the value from open space is higher than the value from development. But of course that might not be so obvious to the developer, or the people who would like to buy houses in the new development, or to other members of the town who care less about the open space than the influx of new residents and the addition to the tax base. Indeed, your own house might once have been part of a new development on what used to be open space. In order to balance the views of people who favor preserving the open space against those in favor of development, we need to have some way of thinking about the benefits from preservation.

How, then, should we think about benefits? From an economic point of view, a person's value for a particular good can be determined by what they would willingly give up in exchange. Economists call this measure of benefits *willingness to pay*. It captures the basic truth that in a world of limited resources, we always have to give up a good thing in order to get another good thing. (We'll come back to this point again when we discuss the rationale behind benefit-cost analysis.) Put differently, if you are not willing to give up anything to preserve the open space, then we might reasonably conclude that your value for it is zero.

While this is a straightforward way of defining "value," several important points should be made. First, by someone's "willingness to pay," we have in mind the amount that the person would just barely be willing to pay. If my

The Costs of Protecting the California Condor

With a wingspan of nine and a half feet, the California condor (*Gymnogyps californianus*) is the largest bird in North America.[1] Until the mid-nineteenth century the condor's range extended as far north as the Columbia River Gorge and south into Baja California. Indeed, the diaries of Meriwether Lewis and William Clark report several sightings of the "Buzzard of the Columbia" in 1805 and 1806. Throughout the twentieth century the wild population declined precipitously, falling from approximately one hundred birds in the 1940s to only nine by 1985. The decline appears to have been due to reduced reproduction (perhaps a result of DDT) and to human-created mortality, including lead poisoning from bullets in game carcasses, shooting of the condors themselves, and hazards from man-made structures such as power lines.

In the late 1980s, the U.S. Fish and Wildlife Service captured the remaining wild birds and embarked on a captive breeding program, with the hopes of eventually reintroducing the species into the wild. In 1992, the first two captive-bred juveniles were released into the Sespe Condor Sanctuary in Los Padres National Forest. By October of 2003, the wild population had climbed to eighty-three birds, including one chick hatched in the wild.

With the condors back in the wild, measures need to be taken to protect the condor populations from threats. From an economic point of view, we can think of these protective measures as "abatement"—in this case, abatement of the causes of condor mortality. Abatement measures include the protection of suitable habitat; provision of food carcasses such as stillborn calves (to prevent lead exposure); promotion of alternatives to lead ammunition; prohibitions on shooting the condors; and modification of power lines and other human structures to reduce injuries to condors.

One study has estimated the costs of abatement using information contained in the Recovery Plan written by the Fish and Wildlife Service. For each abatement action, the number of condors saved per year was estimated taking into account historical rates of decline in condor population and the priority accorded that action by the Service. Unit cost (per condor per year) was then calculated by dividing the cost by condors saved. Arranging the unit costs in increasing order produces a marginal cost function, as illustrated by figure 3.1.

The figure illustrates two key points. First, note the wide range in the marginal costs of various techniques: from as little as seven thousand dollars per condor saved per year to protect habitat in low-lying areas, to over two hundred thousand dollars per condor per year to modify power lines and step up law enforcement. Second, note that it is the marginal cost, rather than the

The Costs of Protecting the California Condor *continued*

total cost, that determines which measures should be pursued first. Thus while the annual cost of heightened law enforcement is only a quarter of the cost of removing contaminants (five thousand versus twenty thousand dollars), contaminant removal would save over thirty times as many condors, and hence is a much more cost-effective means of protecting the species.

Marginal costs of condor protection

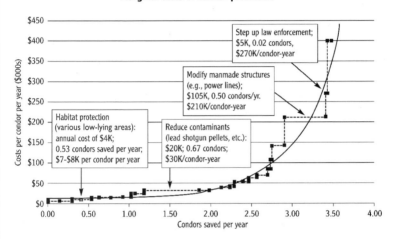

Figure 3.1 Marginal cost graph for condor example. Each "step" on the dashed line corresponds to a specific protection measure, arranged from lowest to highest unit cost. The boxes highlight four specific actions among over two dozen considered. The solid line represents a smooth approximation to the "staircase" function.

willingness to pay to preserve a parcel of open space is one hundred dollars, it means I would be just as happy if I paid one hundred dollars and the open space was preserved as if I paid nothing and the open space was developed. In this sense, willingness-to-pay really measures the *maximum* an individual would be willing to pay.

A person's value for a particular good can be determined by what they would willingly give up in exchange. Economists call this measure of benefits willingness to pay.

Second, nothing in this concept requires that any payment actually be made. My willingness to pay exists as a measure of value, independently of whether I actually write a check for a hundred dollars or not. We are concerned here with willingness to pay "in one's heart of hearts," so to speak.

Third, this concept of value is unabashedly anthropocentric, or human-centered. To economists, the value of anything depends on the satisfaction (or "happiness" or "utility") that humans derive from it. Note that this is not the same thing as saying that only "useful" things have value. On the contrary, the economic notion of willingness-to-pay is as expansive as people's imaginations. A value for open space or biodiversity or endangered species is perfectly consistent with it, since humans have shown (through their actions) that they are willing to sacrifice much to preserve them.

Fourth, in addition to being human-centered, the economic notion of value centers on the preferences of the individual rather than of the group. Economic analysis emphasizes the values and preferences of individuals, *as they perceive them*. One might say that value is in the eye of the beholder. Economic theory is rigorously neutral with respect to what people decide is valuable to them. Most people tend to think of one set of preferences as "right" or "superior" because it seems obvious to them. For example, it might be tempting to think that "everyone ought" to conserve water or gasoline as much as possible. But from the viewpoint of economics, we must respect the preferences of those who would rather keep an enormous lawn green, or drive a gas guzzler. This does not mean that people who prefer to drive Hummers rather than Priuses should not bear the consequences of the extra carbon dioxide, nitrous oxide, and particulate matter they emit. To the contrary, as we shall see in chapter 5, economics prescribes that people (and polluters) pay the full costs to society for their actions. However, the individualistic perspective of economics does hold that people should ultimately be allowed to make their own decisions about what they value and what they do not.

A drawback of this approach is that it does not necessarily take account of how well informed people are about the environment. For example, if people do not understand the ecosystem services that wetlands provide (e.g., flood control, habitat, and water quality) they may place less value on those resources than they would if they knew more. A natural solution is to educate people about the benefits of environmental protection. For example, when economists conducted surveys in the aftermath of the Exxon *Valdez* oil spill in Prince William Sound, they took care to provide some background information on the spill and its likely environmental effects, before asking people to estimate the damages from the spill.

Fifth, a subtle but potentially crucial distinction arises between willingness to pay (WTP) and *willingness to accept* (WTA). The difference between the two hinges on the implicit assignment of property rights. If I ask you what you are willing to pay to preserve open space from development, I implicitly assign a property right to the developer. If, instead, I ask what you are willing to *accept* from the developer in compensation for the very same parcel of land, I have given you the implicit property right. In practice, these implicit property rights are often determined by the status quo. For example, if we contemplate a prospective action that has not yet taken place ("Should we as a society allow drilling in the Arctic National Wildlife Refuge?"), then we might naturally consider society's willingness to accept compensation for loss of the wilderness area. On the other hand, in the aftermath of the Exxon *Valdez* accident, the relevant issue for valuation was society's willingness to pay to clean up the spill. A related way of stating the distinction is that WTA is a natural measure for your losses, while WTP corresponds more readily to gains.

In many cases, WTP and WTA should coincide, at least if we can measure them accurately (i.e., if we can set aside reasons why people might intentionally understate their willingness to pay or overstate their willingness to accept). For example, suppose you head to Fenway Park in Boston to see the Red Sox play a baseball game. Let's assume you would be willing to pay up to (but no more than) fifty dollars for a grandstand ticket. If you show up on the day of the game without a ticket, and find someone willing to sell you one for fifty dollars or less, you will buy it. On the other hand, if you already have a ticket, and someone offers you fifty dollars or more for it, you will sell. If you were not willing to pay eighty dollars to buy a ticket, you will not be willing to forego eighty dollars if you can sell it for that much.

As the stakes get larger and larger, however, a gap may open up between what you would be willing to pay and what you would be willing to accept. This is simply because willingness to pay depends in part on *ability to pay*. To see this, consider the Great Barrier Reef in Australia, which is threatened by the rising ocean temperatures associated with global climate change. How much you would be willing to pay to prevent the reef from being decimated? While the amount might be large, it is constrained by your need to have money to buy other goods, and ultimately by your wealth. Now consider how much you would be willing to accept to allow the reef to be destroyed. Your answer now might well be "infinity," or equivalently "no amount of money." In that case WTA far exceeds WTP. But it helps illustrate an important point. WTA will always be at least as large as WTP. The two measures may diverge considerably, however, when large income effects are involved. Keeping that in mind, we will nonetheless refer to WTP rather than WTA when we consider the benefits of environmental protection.

Finally, the use of willingness to pay as a measure of value is not confined to environmental goods: on the contrary, the concept is fundamental to economic theory. The same principle applies to shoes or coffee or automobiles as to open space: The benefits you derive from a particular good can be measured by what you would be willing to give up for it. Of course, in a market setting, what you would be willing to give up for something can be inferred from your purchase decisions. If you pass up a pair of shoes when they cost one hundred dollars, but buy them on sale for eighty dollars, then I can infer that the value you place on the shoes is between eighty and one hundred dollars. In contrast, no such observable data exist for most environmental amenities. Indeed, as we shall see below, the main practical hurdle to valuing environmental goods is that clean air, endangered species, open space preservation, and other environmental amenities are not traded in markets.

A Taxonomy of Values

Defining the benefits of environmental quality in terms of willingness to pay still leaves us with a range of kinds of values—that is, reasons people might be willing to pay. Let's return to the example of open space preservation. Some people would be willing to pay for open space because they are eager to visit it—to take walks in the woods, or look for birds in a wetland. Other people without current plans to visit the area might nonetheless be willing to pay something to preserve it now, in order to have the option to use it in the future. Still others might not plan on using it themselves, but would want to preserve it to pass on to their children or grandchildren. Finally, some people might be willing to pay something to preserve a parcel of open space simply because they took pleasure from knowing that it existed.

As the example makes clear, we can identify several distinct types of value. Today's visitors value open space because they will use it for recreation, while prospective visitors want to preserve the option of using it in the future. Both of those values fall under the heading of *use value*. Use values involve direct enjoyment or consumption of an environmental good. For another example, consider the gains from reducing smog in Los Angeles. The benefits range from attenuated adverse health effects (eye irritation, asthma, difficulty breathing) to aesthetic values (better views from homes high in the canyons or hiking trails in the mountains nearby). Both the health effects and the aesthetic values constitute use values, since they represent direct consumption or enjoyment of the clean air. Use values also arise from recreation—as for water quality in trout streams, or water volume in river rapids.

The other two types of value mentioned above—the desire to preserve a resource for future generations or the pleasure taken from the knowledge that something exists—are, naturally enough, termed *non-use values*. They involve benefits derived from the existence of an environmental amenity, *but not from its direct use*. Existence value is of particular importance for endangered species preservation in the real world: How many donors to wildlife conservation societies actually believe they will see a Bengal tiger or a polar bear? Rather, much of the value of endangered species or habitats is simply "knowing that they are there."

Measuring Benefits

To apply these concepts, of course—and to carry out benefit-cost analysis—we must be able to measure how much people are actually willing to pay for a given environmental amenity. An in-depth discussion of valuation methods is beyond the scope of this book. However, we can sketch out the basic intuition behind the major approaches.

For most consumer goods, measuring benefits is straightforward. To determine how much consumers are willing to pay for dress shirts, or compact discs, or home appliances, analysts can gather market data on the prices of goods and the quantities purchased. This conceptually simple approach does not work in the case of environmental quality, however, because environmental goods are typically not traded in markets.

Economists have developed two basic strategies to circumvent this lack of price and quantity data in order to estimate the value of environmental amenities. The first approach is to observe behavior in related markets, and use that information to infer willingness to pay for environmental quality. Economists term this the *revealed preference* approach, because it treats actual behavior as "revealing" the true underlying preferences of individuals.

One such method, the *travel-cost method*, infers individual marginal willingness to pay for environmental quality from decisions about where to travel for recreation. For example, if we observe an angler traveling to a remote fishing lake, we can infer that he values the experience of fishing there at least as much as the trip's cost (in time as well as money). By comparing the frequency and duration of visits to pristine lakes with those to relatively polluted ones, an analyst could estimate the marginal benefits of clean water.

Another revealed-preference method uses observed market prices to infer the *implicit* prices for environmental amenities that are bundled with other (private) goods. A common application is to housing markets. Consider the following thought experiment: Imagine two houses located in the same metropolitan area that are identical in every respect, except that one of the houses is in a neighborhood with poorer air quality. The house with cleaner air will command a higher price, with the size of the premium being determined by the difference in air quality and by people's marginal willingness to pay for clean air. In effect, the housing market does set a price on clean air.

Of course, in the real world, houses are never identical on every dimension except for air quality: One is located on a bigger lot and in a better school system, while the other house has a shorter commute and is close to a city park. Instead of being directly observable, the "price of clean air" in the housing market must be teased out from other attributes. Dealing with these other factors, however, is a methodological rather than a conceptual challenge, and sophisticated statistical techniques have been developed to isolate the effect of air quality from other (potentially correlated) factors. Thus the *hedonic housing price method*, as it is called, has been used widely to estimate marginal willingness to pay for reductions in a variety of air pollutants, as well as for other local environmental (dis)amenities, such as toxic waste sites. A similar method can be applied to labor markets to estimate the wage premium workers receive for accepting a high-risk job, or a job in an undesirable location.

The second basic approach to estimating value is simply to ask people how much they would be willing to pay to protect a given environmental resource (e.g., an area of habitat, a population of endangered species, etc.) This *stated preference* approach consists of one basic method (implemented in many different ways) called "contingent valuation" (CV).

CV methods are conceptually straightforward: A group of people are surveyed for information on their willingness-to-pay. The great advantage of this approach is its broad applicability. Revealed preference methods, by their very nature, can be applied only to estimate use values. Existence value is purely internal, rather than behavioral: it leaves behind no "paper trail" of observed behavior that can be used to infer someone's value for a resource. In contrast, CV can in principle be used to measure willingness-to-pay for any good—simply by asking people.

Of course, this broad applicability does not come cheap. The disadvantage of CV methods is that the respondents lack strong incentives to tell the truth. If they feel they have little stake in the outcome—thinking, for example, "this is just a survey"—then they may have little reason to think carefully about their choice and provide a thoughtful response. Things may be even worse if respondents believe that their answers to the survey will affect policy, since in that case they may have strategic incentives to misrepresent their true valuations. For example, suppose I have a moderate willingness to pay for clean air, and live in an area with a large population. If I think that I will be asked to pay a tax based on how I respond to a CV survey, I will have an incentive to understate my true valuation, in the hopes of "free riding" off of the contributions of others. After all, I might reason, my own tax payment is likely to have very little effect on air pollution—but I would be much better off personally if that tax payment were as small as possible. On the other hand, if I believe that my preferences for clean air will help secure strong regulation, but that I will not be asked to pay on the basis of my response, then I may well overstate my willingness to pay, in the knowledge that the cost of clean air will be largely borne by others.

Economists who specialize in contingent valuation have devised a number of approaches to mitigate these and other potential sources of bias. As a result, a range of meta-analyses that have compiled results from multiple studies have concluded that CV methods and revealed-preference approaches yield similar estimates of willingness to pay. This is taken as evidence that neither approach is biased. However, the estimates produced by contingent valuation are generally much more variable across studies. Economists tend to put greater trust in the revealed-preference approaches—just as you might wisely attach more weight to what another person actually does, rather than what she says. Nonetheless, CV remains invaluable for its ability to provide estimates of existence value.

Both the revealed preference and stated preference approaches to estimating the value of environmental amenities tend to focus on *marginal* willingness to pay. Studies that infer value from behavior (such as the hedonic housing price studies) necessarily focus on marginal willingness to pay,

What Is the Value of Global Ecosystem Services? A Cautionary Tale

In a well-known article published in *Nature* magazine, Robert Costanza and a group of colleagues set about estimating the total worldwide value of renewable ecosystem services.[2] They focused on seventeen types of ecosystem services, including pollination, nutrient cycling, and regulation of the composition of gases in the atmosphere, as well as more mundane goods and services produced by nature, such as food production, raw materials (e.g., timber), and recreational opportunities. Their goal was to demonstrate the economic significance of such "natural capital"—most of which is not traded in the economy, and hence has no "market value."

Their conclusion? The study estimated the annual value of such ecosystem services worldwide to be $33 trillion. That might sound like a lot of money. After all, as the authors point out, it is just shy of twice the global GNP at the time of the study. On reflection, however, this amount starts to look fairly small. After all, the world economy grew at about 2 percent per year in 1995. At that rate, the world economy would surpass the value of ecosystem services by the year 2025 or so. The elimination of major global ecosystem services would obviously have devastating effects on human welfare. Indeed, as one analyst quipped, $33 trillion is "a serious underestimate of infinity."[3] Where did the analysis go wrong?

An important part of the answer is that they use estimates of marginal willingness to pay in computing the total values of ecosystem services. Of course, these concepts are closely related, as we saw in chapter 2. But we also saw that marginal willingness to pay often depends on the level of environmental quality. For example, to answer the question "What is the value of pollination services?" Costanza and his colleagues figured out the value of pollination for the marginal hectare, and then multiplied it by the total land area that is pollinated. This is akin to the difference between asking "what is the value of a particular bee colony?" and asking "what is the value of all of the bees in the world?" We can't answer the latter question by scaling up the value of a single beehive. As beehives became scarcer, they would become more valuable.

A related point is that the complete elimination of vital ecosystem services would affect society in fundamental and interconnected ways. When economists measure marginal benefits, they evaluate small changes. The loss of all wild pollinators worldwide—or, even more dramatically, the shutdown of processes that regulate Earth's atmosphere—would cause drastic shifts in the demand for and supply of all kinds of goods and services (including the very amenities and services being measured).

A Cautionary Tale *continued*

Does this mean that the benefit techniques we have discussed are not useful? On the contrary, they are eminently suited to assessing the effect of specific policies, such as reducing concentrations of air pollution, or setting aside a wilderness area for habitat protection. They are simply inadequate to measure of the value of global ecosystem services in their entirety.

---

because that is what can be observed directly. The price of clean air is essentially a measure of what people are willing to pay for an incremental improvement in air quality—just as the price of a CD is the price of a single CD. Contingent valuation studies also tend to focus on marginal willingness to pay—in large part because policy decisions tend to be made on the margin. That is, the relevant question for pollution control policy is not "Should we reduce air pollution?" but rather "Should we reduce pollution by 8 million tons a year, or by 10 million tons?" Similarly, debates over endangered species laws focus on the level of protection that should be provided, rather than on whether endangered species should be protected.

Benefit-Cost Analysis

Thus far, we have discussed efficiency as a "best" or "maximal" outcome—for example, a particular level of pollution control that maximizes net benefits. The concept of efficiency is also useful in comparing alternatives, neither of which necessarily maximizes net benefits relative to all other possible policies. The basic principle is straightforward: Policy A is more efficient than policy B if the net benefits are greater under policy A. For example, Congress may consider whether to require electric power plants to install pollution control devices to reduce mercury emissions into the atmosphere, comparing a proposed regulation to the status quo. A regional development agency may decide whether to spend transportation funds on highway expansion or on construction of a light-rail network. In cases such as these, a systematic comparison of benefits and costs can be a useful aid to government policy.

Critiques of Benefit-Cost Analysis

Although it is simple to describe, benefit-cost analysis has attracted considerable controversy—especially in the environmental arena. Critics of

Benefit-Cost Analysis in the Real World: EPA's Study of Lead in Gasoline

In the fall of 1983, the U.S. Environmental Protection Agency initiated an internal study into the benefits and costs of reducing lead in gasoline.[4] The use of lead as a fuel additive to increase octane levels had been restricted since the early 1970s, but high levels of lead persisted in the environment, partly because older leaded-fueled cars remained on the road, and partly because some owners of newer cars "misfueled" their cars by using leaded gasoline.

The agency's Regulatory Impact Analysis (RIA), published in 1985, identified four main benefits from phasing out lead. Two were linked to reductions in the main adverse human health effects from lead: retardation of cognitive and physiological development in children, and exacerbation of high blood pressure in adult males. A third benefit was a reduction in emissions of other pollutants from cars, since burning leaded gasoline destroyed the effectiveness of catalytic converters. The final benefit was lower costs of engine maintenance and related increases in fuel economy. Meanwhile, the primary costs of phasing out leaded gasoline were the costs of installing new equipment at refineries and of producing alternative additives to boost octane levels.

The study found that the benefits of reducing lead would substantially outweigh the costs. For a ten-fold reduction in lead content (from 1.1 to 0.1 grams per gallon of gasoline), the net benefits were a little over $7 billion dollars annually (in 1983 dollars): $7.8 billion in benefits minus $600 million in increased refining costs. These findings helped support the EPA's decision to accelerate the required removal of lead from gasoline. That reversed the trend set a few years earlier, when the agency—citing costs to refineries—had settled on a much weaker rule than preferred by environmental and public health advocates.

That the benefit-cost analysis succeeded in bolstering the case for tighter regulations was all the more notable because of the gaps in the analysis—gaps readily acknowledged by the study's authors. EPA staff were only able to include the avoided costs of medical care and remedial education for children with blood levels above a certain threshold—leaving out the lion's share of benefits, such as the willingness-to-pay to avoid lasting health and cognitive impacts.

The leaded gasoline example illustrates the usefulness of computing benefits in dollar terms to make them commensurable with other benefits and with costs. Indeed, the case for removing lead from gasoline would have been even stronger had the analysts been able to estimate the dollar value of improving children's health and cognitive abilities. Quantifying benefits (as

---

Benefit-Cost Analysis in the Real World *continued*

well as costs) can focus regulatory efforts on those areas likely to yield the greatest net benefits to society. As one of the study's authors has pointed out, there was little political pressure on EPA at the time to tighten lead standards. Indeed, much more attention was being devoted to issues like hazardous air pollutants and uranium mill tailings, even though the damages from such problems (and thus the benefits from cleanup) were orders of magnitude smaller than those at stake with leaded gasoline.

---

benefit-cost analysis typically advance four main arguments against it.[5] First, basing decisions simply on whether benefits outweigh costs omits important political and moral considerations, such as fundamental rights or duties. Second, discounting benefits that will occur in the distant future privileges current generations at the expense of future ones. Third, goods such as clean air or clean water are devalued and cheapened when their worth is expressed in monetary terms. Finally, focusing on the net benefits to society as a whole ignores the identities of the winners and the losers—that is, an emphasis on efficiency obscures a consideration of distributional equity. We consider each of these criticisms in turn. Doing so allows us to probe more deeply the usefulness of economic efficiency and the limitations of economic analysis.

"Benefit-Cost Analysis Should Not Be the Only Criterion for Decision-Making"

Critics of benefit-cost analysis often contend that economic theory prescribes the use of benefit-cost analysis to the exclusion of other considerations. The emphasis that economists place on the concept of efficiency may initially leave the mistaken impression that they view net benefits as the only criterion needed for public policy. In fact, most economists reject such a narrow view. The consensus among economists is that benefit-cost analysis should be viewed as a means of improving the information available to decision makers—not as the sole guide to decision-making. In the words of a blue-ribbon panel of economists led by Nobel laureate Kenneth Arrow:

> Although formal benefit-cost analysis should not be viewed as either necessary or sufficient for designing sensible public policy, it can pro-

vide an exceptionally useful framework for consistently organizing disparate information, and in this way, it can greatly improve the process and, hence, the outcome of policy analysis.[6]

"Discounting Is Unfair to Future Generations"

A second criticism concerns the use of discounting to compare costs and benefits over time. It may seem grossly unfair to weigh costs to people living today more heavily than benefits to future generations. Indeed, economists recognize full well the thorny issues of intergenerational equity raised by discounting. William Nordhaus has illustrated the dilemma with a particularly striking example. Suppose we discover that Florida will be entirely destroyed two hundred years from now by an asteroid impact. Suppose that we could prevent this catastrophe by launching a missile today to intercept the asteroid. How much should we be willing to pay for the missile? Using the value of land and capital in Florida, and employing the 7 percent discount rate mandated by the U.S. Office of Management and Budget for use in government benefit-cost analyses, Nordhaus calculates that preventing Florida's annihilation two centuries hence would be worth only about $3 million. If launching the missile cost more than that, a strict benefit-cost test (using the government discount rate) would advise against its launch.

This thought experiment helps to highlight the importance of the choice of discount rate. After all, if Nordhaus applied a 2 percent discount rate—putting relatively more weight on the future—the present value of preventing the destruction of Florida would be around $43 *billion*—still small, perhaps, but several orders of magnitude more than the $3 million figure above.

At a deeper level, however, this thought experiment illustrates the limitations of using efficiency as the only criterion for decision making. Sometimes, benefit-cost analysis may suggest a course of action that we might still choose to reject on the basis of ethical considerations—for example, concerns about intergenerational equity. But that does not mean that we should not do the analysis in the first place. In making decisions about how to trade off current costs and future benefits, we as a society are surely better off when we have more information about the choices we face. Carrying out a benefit-cost analysis does not commit us to any particular course of action: it simply helps to clarify the stakes. This argument echoes the one made by the blue-ribbon panel, cited above.

*"Putting Benefits in Dollar Terms Cheapens the Worth of
the Environment"*

A third common criticism of benefit-cost analysis is that it relies on monetary measures of the benefits of environmental amenities such as clean air or endangered species preservation. Critics ask how we can attach monetary values to such "priceless" resources without devaluing them. The problem with this critique is that we need a common yardstick for comparing costs and benefits. Expressing them in dollar terms is often convenient, simply because the costs of implementing a government policy are often naturally expressed in monetary terms—for example, the increase in electricity bills that would result from more stringent pollution controls on power plants.

Of course, a critic might respond that any attempt to make benefits and costs commensurate is problematic, regardless of the metric used. In a world of limited resources, however, trade-offs must be made among competing demands. Weighing costs and benefits is simply a way of assessing those trade-offs. As Nobel laureate Robert Solow has argued,

> Cost-benefit analysis is needed only when society must give up some
> of one good thing in order to get more of another good thing. . . .
> The underlying rationale of cost-benefit analysis is that the cost of the
> good thing to be obtained is precisely the good thing that must or will
> be given up to obtain it.[7]

A related critique holds that there is a moral imperative to protect the environment, making the monetary value of no consequence. A problem with such arguments, however, is that they overlook other conflicting but equally valid appeals to morality. If saving the spotted owl is the morally right thing to do, what about saving the jobs of loggers whose livelihoods depend on cutting down trees? If protecting the Amazonian rainforest from slash-and-burn agriculture is a moral imperative, what about making sure that people have enough to eat? At the very least, expressing benefits and costs in monetary terms can help inform these trade-offs.

To make this concrete, let's consider the control of mercury emissions from electric power plants. Some might argue that the health risks from mercury pollution are so great that mercury should be controlled at all costs. If this argument sounds appealing, ask yourself how much you would be willing to pay in higher electricity bills each month in order to help pay for the costs of installing and operating mercury controls. Would you be willing to pay five dollars a month? Ten dollars? One hundred dollars? One thousand dollars? How would your answer depend on the benefits to you and your family (and society at large) from controlling mercury pollution? Presumably, if the benefits were small enough and the costs were high enough, you might conclude that the emissions controls would not be worthwhile. Spending more money on higher electricity bills to pay for pollution control necessarily entails spending *less* money on something else that you value. It is in this sense that money is a useful measure of benefits— not because we think there ought to be an intrinsic dollar value put on everything, but because trade-offs must be made.

Finally, it is worth pointing out that whether we realize it or not, we end up putting implicit values on environmental quality through the decisions we make every day. If you buy conventionally grown fruit and vegetables rather than the more expensive organic alternatives, you are putting a value on the environment. The same is true if you drive rather than bike to work or to school. The question is not whether we put a value on the environment. Rather, the question is whether or not we make that value explicit.

"Benefit-Cost Analysis Ignores the Losers from a Policy"

A fourth major criticism of benefit-cost analysis is directed more broadly at the notion of using economic efficiency as a criterion for social welfare. As we have seen, efficiency is concerned with the overall net benefits to society from a policy—not with who gains and who loses. For example, the allowance trading program of the 1990 Clean Air Act Amendments reduced sulfur dioxide emissions from electric power plants by roughly 50 percent in the latter half of the 1990s, at a much lower cost than had been expected. (We'll learn more about the trading program in chapter 10, when we discuss market-based environmental policies.) Careful analysis has found that the benefits from that program—principally, reduced mortality and morbidity from air pollution in cities of major power plants—far outweighed the costs of reducing SO_2 emissions.[8]

One contributing factor was the availability of very low-sulfur coal from the Powder River Basin in Wyoming. In the Congressional debates during the run-up to the passage of the amendments, Senator Robert Byrd of West Virginia led the opposition to the new law—largely out of a concern that it would threaten the jobs of miners in his home state. Although the benefits were much larger than the costs in aggregate, West Virginia coal miners

What Happens When Costs and Benefits Are Not Considered Systematically?

A compelling argument for benefit-cost analysis is that when such analysis is not performed, actual policy may reflect implicit biases rather than reasoned considerations.[9] A study of endangered species management by the U. S. Fish and Wildlife Service (USFWS) illustrates this point. Andrew Metrick and Martin Weitzman gathered data on the amount of money spent on endangered species protection, along with data on a variety of species characteristics—including "scientific" attributes such as genetic uniqueness and degree of endangerment, along with "visceral" attributes such as the size of the animal. They found that the USFWS tends to lavish money on protecting animals that are physically appealing, even though they have reasonably large breeding populations and close genetic relatives in no danger of extinction. At the same time, relatively little money is spent on species that face far greater risk of extinction (such as the Monitor Gecko or the Choctawahatchee Beach Mouse) or that are genetically unique (such as the Red Hills Salamander or the Alabama Cave Fish). Indeed, the amount of money spent to protect listed species was strongly and positively correlated with size, but *negatively* correlated with an objective measure of endangerment.

In short, spending by the USFWS on endangered species is biased toward "charismatic megafauna" like grizzly bears, rather than adhering to the scientific principles and priority-setting that ostensibly guide decisions. We are not arguing that society should elevate scientific standards over other reasons people might value endangered species. Rather, the point is that the USFWS is not meeting the criteria it sets for itself as its goals—in part due to an absence of a calculation of the benefits and costs of protecting different species. As the authors conclude in another paper on the topic, good stewardship requires "confronting honestly the core problem of economic tradeoffs—because good stewardship of natural habitats, like almost everything else we want in this world, is subject to budget constraints. The evidence suggests that our actual behavior may not reflect a reasoned cost-benefit calculation."

may well have ended up worse off, as Wyoming coal displaced some of the higher-sulfur Appalachian coal. The costs of sulfur dioxide control also fell heavily on electric utilities in the Midwest, who passed the costs onto their ratepayers in the form of higher electric bills. The benefits from clean air, meanwhile, accrued to residents of downwind cities as well as to anglers and hikers who gained from the reduction in acid rain in the Adirondacks and elsewhere.

A program can be economically efficient, therefore, and still not make everyone better off. In simple terms, efficiency is about "maximizing the size of the pie," while distributional equity is about who gets what share of the pie. This potential conflict between efficiency and distributional equity is fundamental—and it is an issue that merits careful consideration in each instance. If efficiency ignores something as critical as distributional equity, you may ask, why should we use it as a benchmark for policy? The answer is twofold: Efficiency provides a welfare criterion that is both fundamentally sound and implementable in practice.

The philosophical foundations of efficiency date back to the work of the Italian economist Vilfredo Pareto at the end of the nineteenth century. Pareto proposed that one policy is superior to another if at least one person is strictly better off under the first policy, and nobody is worse off. A policy is "Pareto efficient" if—and only if—no member of society could be made better off by an alternative policy without making at least one person worse off.

As a criterion for comparing a proposed policy with the status quo, Pareto efficiency has a certain appeal. If policy A makes at least one person better off than policy B, without harming anyone, would we not always want to adopt policy A?[10] When we try to apply this criterion in the real world, however, a drawback becomes clear: It is much too strict. Using it as a guide to making policy would almost always favor the status quo. Think of nearly any medical or technological breakthrough which was undoubtedly beneficial to society as a whole, and you can come up with at least one group of people who was hurt by the change. The rapid expansion of the railroad in the second half of the nineteenth century spurred the development of the western United States, brought fresh meat and produce to urban populations, and led to the rise of Chicago as a great city; but it also put bargemen out of business on the now-obsolete canals and contributed to the overexploitation of natural resources such as the American bison. The development of the personal computer, for all its undeniable benefits, also devalued the skills of professional typists and shuttered countless old typewriter stores.

How to reconcile the obvious benefits to society from such changes with their smaller but very real costs? A way out of this conundrum was proposed separately by Nicholas Kaldor and John Hicks in the 1930s. Under the Kaldor-Hicks or "potential Pareto" criterion, policy A is chosen over policy

B if policy A would make at least one person better off without making anyone worse off, *provided that suitable transfers were made from the winners to the losers*. Crucially, the Kaldor-Hicks criterion does not require that those transfers are actually carried out. In other words, the modified criterion can be thought of as satisfying the strict Pareto criterion in principle, given appropriate (but possibly unfulfilled) transfers. Of course, compensating the losers from a given policy is much easier to describe than to carry out. Nonetheless, it remains crucial that such transfers be possible, even if only in theory—because the possibility of such compensation ensures that there is a net surplus from the policy. The winners from a policy can only compensate the losers (even in principle) if the benefits from the policy are greater than the losses. Hence a policy satisfies the modified criterion if and only if it produces greater net benefits than the alternative. Satisfying the Kaldor-Hicks criterion is equivalent to maximizing net benefits.[11]

A policy is "Pareto efficient" if—and only if—no member of society could be made better off by an alternative policy without making at least one person worse off.

The Kaldor-Hicks approach also helps clarify the relationship between efficiency and distributional equity. Since the compensating transfers need not actually take place, an efficient policy may lead to one group gaining much at the expense of another. If we are interested in distributional equity, this indifference to whether the transfers occur is a crucial flaw.

A policy satisfies the "potential Pareto cirterion" if it would be Pareto efficient provided that the winners from the policy compensated the loser—but such transfers do not actually have to take place.

Three main responses are possible to such a critique. First, one can simply point out that efficiency and distributional equity are competing goals; both deserve consideration when shaping environmental policy, or policy in any other sphere. Second, we might argue that over time and across a broad range of policies, the gains and losses enjoyed by any particular group of people in any particular case tend to cancel each other out—so that in the long run, pursuing efficient policies will improve everyone's lot. If today's winners are tomorrow's losers, then we need not concern ourselves too much with the distributional implications of particular policies. However, such a sanguine view of affairs is clearly naïve when it comes to the impacts of much public policy in market economies like the United States, where in the absence of redistributive programs the pursuit of economic efficiency is likely to further concentrate wealth in the hands of a few.

A third response to the critique is to find creative ways to spread the gains from efficient policies. A real-world example of just such a compensation program is the Northwest Forest Plan enacted in 1993 by the Clinton administration. The forest plan grew out of concern over declining populations of northern spotted owls and the old growth habitat they depended on. To protect the owl and its habitat, the plan sharply reduced the allowable timber harvest on 24.5 million acres of federally owned lands in Washington, Oregon, and northern California. Loggers in the local timber industry were among the most vocal opponents of such conservation measures; pickups and logging trucks sported bumper stickers reading "Save a Logger: Eat a Spotted Owl" and "Spotted Owl Tastes Like Chicken." In response to such concerns, the Clinton Administration proposed a Northwest Economic Adjustment Initiative. That program provided $1.2 billion over six years in additional funding to cushion the blow to logging communities through job retraining, rural development assistance, and direct payments.

Conclusion

This chapter has tried to flesh out the concept of economic efficiency that we developed in chapter 2. In order to put that into practice, we need to be able to measure costs and benefits. Costs are more straightforward to define and to measure. Even so, economics raises two key insights: The costs of doing one thing depend on the opportunities foregone, and the burden is often shared by many groups in society. The benefits from environmental amenities—as from any other good—ultimately derive from what people are willing to pay to secure them. Economists have devised a range of clever approaches to infer such willingness to pay from how people behave when they go on vacation, buy a house, or accept a job. Nonetheless, in some cases—for example, the existence of an endangered species—there is no alternative but to ask people how much they value a good.

Even if we can measure costs and benefits, it is important to have a firm grasp on why economic efficiency is a desirable goal for society. As a criterion for policy making, efficiency has two important strengths: It is grounded in a strong welfare justification (recall the Pareto criterion it is based on), but at the same time is readily implemented in practice. Nonetheless, there are powerful critiques of efficiency, and of the intertwined practice of benefit-cost analysis. In particular, a single-minded focus on efficiency can mask deeply unfair outcomes if the positive net benefits reflect large gains to one group of people at the expense of another. Keeping this blind spot in mind, however, we shall use efficiency throughout the remainder of the book as our basic benchmark for evaluating environmental policy and the management of natural resources.

4

The Efficiency of Markets

How well do market economies deal with environmental problems? One school of thought holds that government regulation is costly and intrusive, hindering innovation and economic growth. The conservative talk radio host Rush Limbaugh captured this sentiment when he wrote, "The key to cleaning up the environment is unfettered free enterprise . . . . Capitalism is good for people AND for other living things."[1] On the opposite side of the political spectrum, observers such as Amory Lovins and Paul Hawken claim that firms can cut costs and boost profits by "being green"—for example, by reducing pollution or increasing energy efficiency.[2] If such a connection between corporate profits and environmental protection were to hold in general, it would suggest that government regulation is unnecessary—since firms would find it in their own self-interest to protect the environment.

So, how well would the free market perform on its own? In this chapter, we discuss the advantages of well-functioning markets. Under certain conditions, market outcomes are Pareto efficient. In other words, markets allocate the production and consumption of goods in a way that maximizes the net benefits to society. This result is evoked by Adam Smith's famous image of the "invisible hand." Without any explicit coordination, the interactions of individual consumers and producers, each motivated by self-interest, nonetheless combine to advance the common good. In general, therefore, free markets are a socially desirable means of allocating goods and services. But not always: In some important cases, markets can "fail." In the next chapter, we shall discuss several closely related notions of market failure that are ubiquitous in the environmental realm. As we shall see, competitive markets—although typically effective institutions for allocating resources—are unlikely to provide adequate levels of environmental quality without some government intervention.

Before diving into our discussion of markets and market failure, one point must be emphasized. While market economies may give rise to environmental problems, the capitalist system should not be construed as the sole *cause* of those problems. The scale of environmental problems in capitalist economies like the United States often pales next to the devastation wrought in socialist economies like the former Soviet Union. We focus here on how environmental problems arise in market economies—not because such problems arise only in market economies, but because markets are the dominant form of economic organization in the world today.

Competitive Market Equilibrium

Let's start by defining what markets are and describing how they work.

Defining Markets

In everyday activity we take part in a range of different kinds of markets: retail stores such as supermarkets or clothing shops, farmers' markets where local farms sell produce from individual stands, open-air flea markets, or capital markets such as the New York Stock Exchange. More broadly, the United States and other western countries have decentralized market economies in which the price and quantity of goods are determined by the interaction of supply and demand, rather than by the central state government.

Markets are not the only way to allocate goods or services. An obvious counterexample is a planned economy, such as that of the Soviet Union during most of the twentieth century. To a lesser degree, state-owned enterprises (which are typically insulated from market forces by subsidies and regulatory control) are common in Europe. But much more broadly, even in a market economy a range of other allocation mechanisms exist: you might purchase antique furniture at an auction; bid for collectibles on eBay; download music from an online common-pool system such as BearShare or Napster; win a prize in a raffle or lottery; receive food stamps from government agencies; gain admission to a competitive college or university through the admissions process; or secure a prestigious medical residency through the medical matching system.

What, then, are the key characteristics of a market?

A market is a decentralized collection of buyers and sellers whose interactions determine the allocation of a good or set of goods through exchange.

Note several points from the definition. First, a market is an institution for allocating goods from those who produce or own them to those who want to buy them. Of course, markets are not the only means of allocating goods, as the examples in the preceding paragraph demonstrate. Second, markets are based on the *exchange* of payment for goods or services, rather than a one-way allocation of scarce resources (as in the cases of food stamps, a lottery, or the medical match system). Finally, and perhaps most importantly, markets are *decentralized*. This crucial characteristic distinguishes them from auctions (where buyers and sellers are brought together but exchange is arranged by the auctioneer) and from centralized economies (where a government planner orders producers to make specified quantities for sale or distribution to consumers).

Demand and Supply

Even if you have never studied economics, you probably know the bedrock of microeconomics: supply and demand. To understand market outcomes, we need to describe the behavior of sellers and buyers. First, consider buyers, who make up the demand side of a market. As the price of a good falls, some people who were already buying the good decide to consume more, while some people who chose not to buy the good at higher prices start to purchase it. In other words, as the price falls, the quantity demanded by consumers rises. This relationship between price and quantity, sometimes called the "law of demand," holds for almost all goods.

We can describe consumers' behavior by a *demand curve*. A demand curve summarizes how much buyers in the aggregate will buy at a given market price, holding all other factors (such as the prices of other goods) constant. The underlying relationship between price and quantity can also be stated another way. At each quantity, the demand curve summarizes what buyers are willing to pay for one more unit of a good, given how much they have consumed already.

Figure 4.1 depicts a hypothetical demand curve for coffee. Note that quantity is on the horizontal axis and price on the vertical axis. The demand curve slopes downward, since lower prices lead to larger quantities demanded. For the sake of discussion, let's imagine that this represents the demand for gourmet coffee in a college town.

Now consider the sellers of a good—the supply side of the gourmet coffee market. A *supply curve* summarizes the relationship between price and quantity on the supply side. (See figure 4.1.) The supply curve represents how much the coffee shops in your town (collectively) are willing to sell or produce at a given price. Equivalently, for any given quantity, the supply curve represents the amount the coffee shops are willing to accept in order to produce one more unit of a good—one more cup of coffee. Note that the supply curve slopes upward. As the price of a good increases, so does the quantity firms are willing to supply. At higher prices, existing producers can expand their output in a range of ways: hiring more workers (perhaps increasing wages to do so); paying higher prices to secure larger quantities of needed inputs, say coffee beans; running their shops

A demand curve summarizes how much buyers in the aggregate will buy at a given market price, holding all other factors constant. A supply curve summarizes the relationship between price and quantity for the sellers of a good.

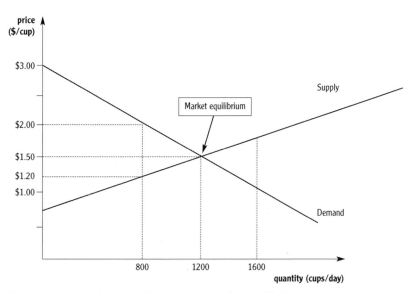

Figure 4.1 Demand and supply curves, and market equilibrium.

more intensely, incurring higher costs; or perhaps even opening new shops. Moreover, higher prices allow new firms with higher costs of production to enter the market, expanding output further.

Market Equilibrium

What happens when supply and demand interact? The answer is that the market will eventually settle where demand equals supply. This combination of quantity and price is called the *market equilibrium*. The term "equilibrium," originally from physics, connotes a stable outcome. Starting at any other quantity, the market will tend toward the price and quantity where demand equals supply; once it reaches this point, the market will stay there, as long as the underlying factors that drive demand and supply (such as income, people's tastes, and production costs) remain unchanged.

In Figure 4.1, supply and demand are equal at a price of $1.50 and a quantity of 1200 cups/day. This point—and only this point—is a market equilibrium. To see why, suppose that instead of producing 1200, suppliers produced only 800 cups. Why can't this be an equilibrium? The reason is that at a quantity of 800 cups, consumers are willing to pay more than the minimum suppliers are willing to accept. In particular, from the demand curve we know that if there were only eight hundred cups available, consumers would willing to pay two dollars a cup. But suppliers would be willing to sell that quantity of coffee for as little as $1.20 a cup: that is what the supply curve reveals. Hence, at least some producers would take advantage of consumers' willingness to pay more and would increase production, pushing the price down. That is true as long as quantity is less than 1200.

Now consider a quantity greater than 1200, such as 1600. Such a quantity cannot be an equilibrium either. Once again, there is a gap between what producers would be willing to accept as a price and what consumers are willing to pay. To produce 1600, suppliers would need to receive a price of $1.80 per cup. (You can see on the figure that the height of the supply curve is $1.80 at a quantity of 1600.) But at this quantity consumers are only willing to pay one dollar. As a result, there would be a surplus, relative to what consumers are willing to buy. Seeing that demand was not strong enough to sustain a high price of $1.80 per cup, suppliers would scale back production until the equilibrium was reached.

The market equilibrium is the combination of quantity and price for which demand equals supply.

Only where supply equals demand—in this case, at 1200 cups/week—can we find a price at which the quantity demanded by consumers equals the quantity supplied by producers. In other words, the market "clears"—leading us to refer to the equilibrium price ($1.50) as the "market-clearing price."

The Efficiency of Competitive Markets

So far, we have described the workings of the market mechanistically—that is, we have predicted what will happen in a particular market, given the behavior of buyers and sellers (as represented by the demand and supply curves). But we've also told you that under certain conditions, the market outcome maximizes net benefits to society. That is, as a general matter, markets are Pareto efficient. How does this happen?

Demand and the Benefits to Consumers

Suppose you would be willing to pay up to $350 for a particular CD player—in other words, you would be indifferent between paying $350 and getting the CD player, or not having the CD player at all. If you find that model of CD player on sale for $299 and buy it, you have in effect received a surplus of $51—the difference between what you would have been willing to pay, and what you actually paid.

The same principle is at work in markets in general. Remember that a demand curve traces out consumers' willingness to pay. Since the market price is determined at the margin, and demand curves slope downward, it will generally be the case that almost all consumers who make purchases in a market pay less for the good than they would be willing to pay. For example, in figure 4.1, the price of coffee is $1.50 per cup, but some consumers would pay as much as $3 a cup. Economists call the resulting benefits consumer surplus: the difference between what consumers would be willing to pay for a good (the area under the demand curve) and what they actually pay (price times quantity). Graphically, consumer surplus corresponds to the shaded area below the demand curve and above the price. Figure 4.2 provides an illustration, for a market in which the equilibrium quantity is $Q^{\star}$ and the price is $P^{\star}$.

Note that we are in effect using the height of the demand curve as a measure of the marginal benefit to consumers. That should make sense. After all, we defined benefits in chapter 3 as measured by willingness to pay. While the discussion of that chapter was focused on environmental amenities, the

> *The difference between what consumers would be willing to pay for a good and what they actually pay is called "consumer surplus."*

same principle applies to all economic goods. From an economic perspective, the marginal benefit an individual receives from consuming a particular good or service can be measured by her marginal willingness to pay for that good or service.[3]

Supply and the Costs to Producers

Next, consider the supply curve. Recall that we defined it as representing the relationship between the quantity of a good produced and the price producers are willing to accept to produce one more unit.

It turns out that the supply curve traces the *marginal cost curve* of the industry. Marginal cost refers to the cost of producing one more unit of a good.[4] Suppose you are in charge of a firm that produces widgets and sells them for five dollars each. If the cost of producing an additional widget is less than five dollars, you can make money by increasing production. For example, if the marginal cost of making a widget is two dollars, you will earn a net return of three dollars (five dollars minus two dollars) from making another widget. On the other hand, if the marginal cost is greater than the price, then you should not produce any more.

Now suppose that your marginal cost increases with the number of widgets you make. (That will usually be the case, at least in the short run when the size of the factory is fixed: as output rises, it becomes more and more costly to increase production, as firms near the limit of their available capacity.) How many widgets do you want to make? The answer is: Produce widgets up to the point where the marginal cost equals the price. As long as marginal cost is less than the price, increasing output will raise revenues by more than cost; the opposite occurs once marginal cost exceeds the price. The same reasoning can be applied at the level of an industry made up of many firms. Firms in the aggregate will produce output at the level where price equals marginal cost. Therefore, the supply curve runs along the marginal cost curve.

In figure 4.2, we have shaded the area below the price and above the supply curve. This area is analogous to the measure of consumer surplus, and—not surprisingly—is termed "producer surplus." What does producer surplus represent, in economic terms? Recall from the discussion in chapter 2 that the area under a marginal cost curve corresponds to total costs. The same

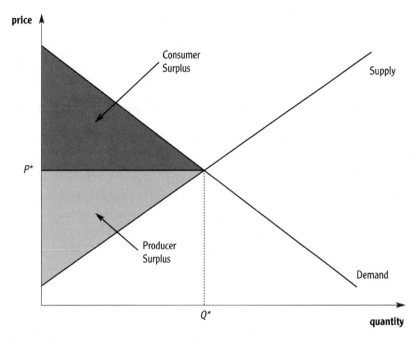

Figure 4.2 Demand, supply, market equilibrium, and welfare measures.

principle applies here. As a result, the area under the supply curve corresponds to the costs of production. Meanwhile, the entire area under the price line equals price times quantity, or total revenues. Thus producer surplus equals the difference between the revenues to suppliers and their cost of production—that is, net revenue to producers.

Market Efficiency

The sum of producer and consumer surplus is a measure of the net benefits produced by market interactions. A careful look at figure 4.2 should convince you that this measure of net benefits is maximized at the point where supply equals demand. Thus the market equilibrium achieves the maximum possible net surplus from the market.

Indeed, we can see this efficiency result as a direct application of the equimarginal rule we discussed in chapter 3. We have already seen that the market equilibrium is given by the intersection of the demand and supply curves. We just argued that those curves can be thought of as, respectively, marginal benefit and marginal cost curves. This suggests that market outcomes equate marginal benefit with marginal cost, and hence must be efficient.

This is an extremely powerful result, but the efficiency of markets is not guaranteed in all cases.[5] In particular, it depends on three necessary conditions. First, markets must be competitive, in the sense that all firms and consumers must take prices as given. That is, individual firms and consumers must be unable to manipulate the market price in their favor. This is a reasonable assumption in many settings where there are many firms or producers competing with each other to sell a good to a large number of potential buyers. When this assumption fails, however, market efficiency fails with it. For example, a single firm with a monopoly over a product will have the power to set its own prices, rather than allow them to be dictated by the market. Rather than set a price that just covers the incremental cost of production, therefore, the monopolist will set a higher price that maximizes its profit. This higher price means that less is sold than in a competitive market, with correspondingly lower surplus to consumers. For example, Microsoft has a near monopoly on operating systems for personal computers, allowing it to set prices for its software far above the marginal cost of production.

The second condition that must be met for markets to be efficient concerns the information available to firms and consumers about the quality of the good or service being traded. A classic example is the used-car market, in which the seller of a car typically knows much more about its quality than any prospective buyer. In such cases, sellers have an incentive to take advantage of their private information; knowing this, wary buyers are likely to be less willing to pay for goods or services whose quality is uncertain.

Finally, for markets to be efficient, they must be *complete*, that is, they must capture all the good and ill effects resulting from a market transaction. In particular, the costs of a good or service must be fully paid for by those who produce it, while the consumer who buys that good or service must enjoy the entire benefit from it. When this condition fails, demand and supply no longer reflect marginal benefit and cost, since part of the benefits or costs accrues to other consumers or firms.

Our market efficiency result can be formally stated as follows:

A market equilibrium is efficient if the following conditions are met:

1. The market is competitive, meaning that firms and consumers take prices as given;

2. Firms and consumers both have good information about the quality of the good or service being traded; and

3. The market is "complete," in the sense that all relevant costs and benefits are borne by the market participants (the firms and consumers involved in transactions).

In many real-world markets these three conditions are likely to hold, at least approximately. While market power certainly arises in the real world, many markets are reasonably competitive. For example, retailers such as supermarkets, bookstores, and clothing stores operate in fairly competitive industries. If your local Stop and Shop supermarket raised the price of bananas or cereal too far above prevailing prices, consumers would simply buy those items elsewhere. If Barnes & Noble charges more for books at its online site, consumers will switch to Amazon. Similarly, consumers and firms typically have adequate (or at least symmetric) information about the quality of goods and services; and when information asymmetries do crop up, market solutions often arise as well, as when car dealerships certify used cars. In general, therefore, free markets are a socially desirable means of allocating goods and services.

When any of these three conditions are not met, however, markets are no longer efficient. In the language of economics, they are said to "fail." Failures of the first two conditions are the focus of extensive literature in economics, and much debate in the real world (witness the controversy about the U.S. government's campaign against Microsoft for alleged antitrust violations). When it comes to the environmental realm, the problem most often is with the third condition—that is, the complete markets assumption. The resulting market failures are the subject of the next chapter.

Conclusion

This chapter has explained how supply and demand interact in a market. The key result that emerges from our discussion is that competitive markets are efficient, at least in many cases. When buyers and sellers take prices as given and have good information about the quality of goods and services, and when the entire costs and

In general, free markets are a socially desirable means of allocating goods and services. When the key conditions for market efficiency are not met, however, a "market failure" is said to occur.

benefits are borne or enjoyed by consumers and producers, the decentralized interaction of self-interested individuals leads to socially desirable outcomes. This is a powerful result, and demonstrating it stands as one of the crowning achievements of economics. Nonetheless, although it is quite general, it does not apply in every case. In particular, as we shall see in the next chapter, market failure is prevalent in the environmental realm, since the costs and benefits of environmental protection and sound resource management are often left out of the calculations of individuals and firms.

5

Market Failures in the Environmental Realm

We've seen that competitive markets are often efficient. If that were the end of the story, this would be a much shorter book (and we, as environmental economists, would have much less to think about). Of course, when it comes to the environment, the assumptions underlying the efficiency of markets commonly fail to hold. Economic activity often gives rise to unwanted by-products, such as water and air pollution, that impose indirect costs on consumers or firms downstream or downwind. And these costs are not usually captured in markets. For example, car exhaust is a major contributor to smog. But while drivers pay for gasoline, tires, maintenance, and so on, they do not pay for the soot and sulfur they send into the atmosphere.

Economists call this sort of thing a *negative externality*. By driving their cars, people impose a cost onto others in the form of poor air quality. That cost, however, is invisible to the drivers themselves; in the parlance of economics it is *external* to their decision-making. Note the asymmetry between benefits and costs. If you drive to work, you gain the entire benefit from driving—in terms of convenience, comfort, and so on—but you share the costs of greater pollution with everyone else around you.

The problem of incomplete markets can also arise in other ways. Some environmental amenities, such as biodiversity, are enjoyed by lots of people, whether or not those people help pay for them. Economists call such goods *public goods*. A market failure arises because some individuals will end up being free riders: Rather than helping to provide the public good themselves, they merely enjoy what others provide for them.

A third class of environmental problems is known as the *tragedy of the commons*. When a natural resource—such as a fishery or an underground aquifer—is made available to all, individuals will tend to exploit the resource far beyond the optimal level. This problem arises because the incentives of individuals diverge from the common good.[1] We call it a tragedy because everyone would be better off if they could all commit themselves to act less selfishly. Thus individually rational actions add up to a socially undesirable outcome.

In this chapter, we examine each of these three problems in turn. We demonstrate why markets fail to be efficient in each case. We also show how these seemingly distinct categories of environmental problems are linked at a fundamental level. You may already see the deep similarities among them. For example, air quality can be described in terms of a negative externality (your automobile exhaust makes my air worse), as a public good (clean air is enjoyed by all, hence individuals have too little incentive to provide it), or as a commons problem (each driver overuses the shared atmospheric commons). Similarly, overfishing problems, while typically couched in the language of the tragedy of the commons, can also be described as a negative externality or as a free-rider problem.

Externalities

We gave an intuitive definition of a negative externality above, in the example of automobile emissions. More generally, we can define an *externality* as follows:[2]

> An externality results when the actions of one individual (or firm) have a direct, unintentional, and uncompensated effect on the well-being of other individuals or the profits of other firms.

Note three key words in the definition: direct, unintentional, and uncompensated. For example, since your health and happiness depend in part on how clean the air is, automobile drivers have a direct effect on your well-being. "Unintentional" is included in the definition to rule out acts of spite or malice. (It is the *effect* rather than the action that is unintentional. I may decide deliberately to use a gasoline-powered lawnmower, without the intent of my action being to pollute the air or disturb the neighbors.) Finally, "uncompensated" implies that the responsible actor is not compensated (or fined) for their actions. This rules out market transactions or bargaining between individuals.

Second-hand cigarette smoke is a common example of an externality: a smoker in a bar ignores the effects of her smoking on nearby patrons (except for her companions—which is why smokers turn away from their friends and blow smoke over their shoulders). At a larger scale, air pollution from factories or power plants represents an externality for downwind populations. Externalities can also impose costs on other firms, rather than individuals. In the Pacific Northwest, logging in forested headwaters degrades spawning habitat for salmon, while hydroelectric dams hinder the fish on their way upstream. Both activities adversely affect commercial fishermen.

While environmental problems are typically framed as negative (harmful) externalities, positive (beneficial) externalities are also possible. For example, a firm that carries out research and development often produces knowledge that its rivals can use—a positive externality, since some of the benefits from the research are captured by firms who do not contribute to its expense. If my neighbors keep their houses and flower gardens well maintained, the value of my house is likely to rise; thus I benefit from their actions, but they do not reap that additional gain. (We will return to the topic of such positive externalities when we consider public goods in the next section.)

How Do Externalities Cause Market Failure?

To see why the efficiency of markets breaks down in the presence of an externality, consider an oversimplified version of the steel industry. Steel furnaces typically burn coal, emitting sulfur dioxide, nitrous oxides, and particulate matter. Suppose for simplicity that there is a fixed relationship between the amount of steel produced and the amount of pollution emitted: for example, for every thousand tons of steel, one ton of sulfur dioxide is emitted. (This would be true if steel mills were unable to install pollution control devices or switch to cleaner fuels, and thus could reduce pollution only by reducing output. We will relax this assumption when we consider policy instruments in chapter 8, but for now it proves useful.)

Now consider the marginal damages of pollution as a function of the amount of steel produced. In figure 5.1, we have drawn such a function (labeled *MD* for marginal damages), along with the familiar supply and demand curves. In the absence of regulation, each steel producer will (quite rationally) ignore the damages caused by its pollution when deciding how much to produce, and consider only its own costs. Indeed, this explains the term *externality*: The damages from pollution are "external" to the firm. The supply curve, therefore, corresponds to the private marginal costs of steel production: the costs of the labor, fuel, materials, and so on required to make one more unit of steel. (We have labeled the curve *PMC* on the graph, for "private marginal costs.") This is exactly the same kind of supply curve we saw in chapter 4. But now there is another cost of production, which the steel producers do not pay—namely, the damages from pollution. What is the efficient level of steel production, given this externality? As always, efficiency means maximizing net benefits to society as whole, and requires equating marginal benefit and marginal cost. Just as in the previous chapter, marginal benefit corresponds to the demand curve. The externality, however, means that the *social* marginal cost is no longer equal to the supply curve, which reflects only the private marginal cost. Instead, social marginal cost (labeled *SMC* in the figure) equals the private marginal cost (paid by the steel industry) plus the marginal damage from pollution. The efficient quantity, where Demand = *SMC*, is labeled $Q^\star$ on the graph.

Now, let's compare that efficient outcome to the one that would result in a free market. Since the supply curve is unaffected by the pollution damages, so too is the market outcome. The unregulated market equilibrium occurs where supply equals demand—the quantity labeled Q^M on the figure. Note that $Q^\star < Q^M$: the unregulated market equilibrium results in too much output (and thus too much pollution). In the absence of regulation, the market yields too much of a bad thing.

In the previous chapter, we saw that the competitive market outcome is efficient when there are no externalities. In the presence of an externality, that result no longer holds. The divergence between the market equilibrium and the efficient outcome arises precisely because the steel producers do not bear the full costs of production. If they paid the marginal damages from each unit of pollution, then their private costs would coincide with social costs, and the supply curve would trace out the *SMC* curve rather than the *PMC* curve in figure 5.1. In that case, the pollution damages would be internalized, and the market outcome would once again be efficient.

To understand the consequences of the market failure, consider what happens when we correct it. Suppose we move from the unregulated market outcome (Q^M) to the efficient outcome ($Q^\star$). The amount of steel produced falls, while the price paid by consumers rises. Therefore, reducing output lowers the sum of consumer and producer surplus. (This must be the case, since it remains true that the market equilibrium maximizes their sum.) The value of this lost surplus is illustrated by the lower shaded triangle in figure 5.1.

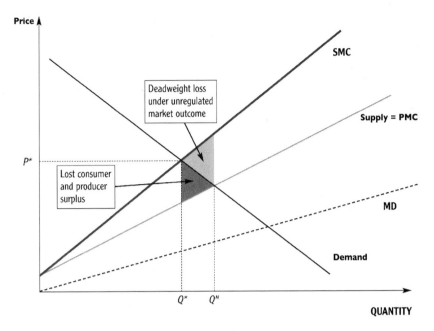

Figure 5.1 A market with a negative externality.

How, then, can moving to Q^* raise welfare, as measured by the total surplus to society? The answer is that social surplus now has an additional component: it equals the sum of consumer and producer surplus, minus the damages from pollution. In a sense, there is another segment of society that must now be accounted for: namely, people who suffer from pollution caused by steel mills. Although curtailing output hurts consumers and producers of steel, it benefits people who are harmed by pollution. Starting at Q^M, the gain from reducing pollution damages outweighs the lost consumer and producer surplus, on the margin. That remains true all the way until output falls to the efficient point, Q^*, where marginal social cost equals marginal benefit.

The upper shaded triangle in figure 5.1 equals the difference between the avoided pollution damages and the lost consumer and producer surplus. Thus it represents the net gain from reducing output from the unregulated level, Q^M, to the efficient level, Q^*. Equivalently, we can think of the same triangle as the net loss to society that results from the unregulated market, relative to the efficient outcome. Economists call this kind of vanished social welfare "deadweight loss"; the name underscores the notion that such losses are not transfers from one group to another, but rather a loss to society as a whole. The deadweight loss, in this case, represents the lost social surplus due to the overproduction of steel (and pollution) in the unregulated market. Its size depends on the slopes of demand and supply, and on the magnitude of pollution damages. Like consumer and producer surplus, the deadweight loss can be expressed in monetary terms: with knowledge of the slopes of the curves in figure 5.1, we could attach a dollar value to the size of the inefficiency represented by the deadweight loss triangle.

Public Goods

A second type of market failure in the environmental realm arises with public goods—goods that are shared by all and owned by no one. National defense is a classic example of a public good. A country's armed forces offer protection from invasion to the citizens living within the country's borders. Importantly, all citizens within the same country are afforded the same level of protection; further, the security enjoyed or "consumed" by one citizen does not diminish that of her neighbor. Biodiversity is a leading example of an environmental public good. Greater genetic diversity makes the food supply more robust to threats from parasites and disease and offers the potential for new medicines or industrially useful chemicals. Many people value the existence of rare or exotic species of animals and plants, or of uninhabited wild places—even if they will never walk in those wild places, or glimpse those animals and plants. Like national defense, everyone enjoys these benefits of biodiversity, and no one person's enjoyment reduces the amount available to others.

These public goods have two fundamental characteristics in common. First, public goods are *nonrival*: the amount of any individual's consumption does not diminish the amount available for others. Second, they are *nonexcludable*: individuals cannot be prevented from enjoying a public good. In particular, even individuals who did not help to provide the public good can still benefit from it. (You breathe the same air whether you drive a Ford Expedition to work and a John Deere lawnmower on the weekends, or ride a bicycle and cut the grass with a

The "deadweight loss" from a policy refers to lost social surplus—not transfers from one group to another, but rather a loss to society as a whole.

push-mower.) Any good that is non-rival and nonexcludable is, by defini-tion, a public good. Of course, a particular good might be relevant only to a given region: Cleaner air in Den-ver is not a public good in Dallas. Sim-ilarly, some goods are "public," but only for a limited and well-defined population: For example, a city park may be open to all city residents free of charge, but not to outsiders.

A type of market failure in the environmental realm arises with public goods—goods that are shared by all and owned by no-one.

We can represent these two characteristics in a box diagram (figure 5.2). The vertical axis measures nonexcludability; the horizontal axis, nonrivalry. Pure public goods (like the examples mentioned above) are located in the upper right corner of the box: they are both nonexcludable and nonrival. On the other hand, pure *private* goods—those that are fully rival and ex-cludable—are in the opposite corner. A candy bar is a simple but illustrative example: If I have a candy bar, I can completely exclude you from enjoy-ing it; and whatever I eat leaves less for everyone else. Most goods com-monly traded in markets—shoes, clothes, furniture, and so on—are purely (or nearly so) private goods.

In between these extremes of pure public or private goods, we can think of goods as having varying degrees of "public-goodedness." Some goods ex-hibit one characteristic but not the other. For example, cable TV is highly nonrival: under most circumstances, the quality of the signal does not di-minish appreciably with the number of users. However, cable TV is perfectly excludable. In contrast to broadcast TV, the cable company can shut off the signal to a consumer who fails to pay. Thus cable TV occupies the lower right-hand corner of our box in figure 5.2. (Because congestion is possi-ble, even if infrequent, cable TV is not all the way at the far right-hand side of the box.) Goods such as this one that are nonrival but excludable are known as "club goods."

At the other corner of our diagram we might put an open-access re-source, like a fishery or forest that is open to all. In these cases, the good is fully rival: If I harvest a tree, it reduces the timber left for you. But by def-inition an open-access resource is nonexcludable. Note that in these cases (as in others) nonexcludability is not an inherent characteristic of the good, but rather is a product of institutions. For example, contrast a freeway (such as an interstate highway) with a toll road: by institutional design, the freeway—

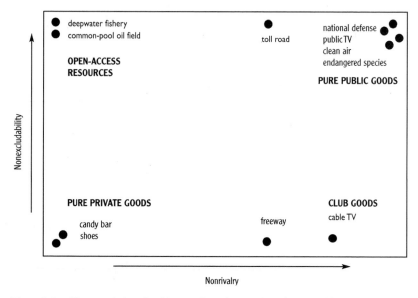

Figure 5.2 Characteristics of public goods and examples of pure public goods, pure private goods, and mixtures.

but not the toll road—is nonexcludable. Or compare a major artery leading into a city at 8 a.m. to the same highway a few hours later. At rush hour, the highway is a rival good: my presence on the road lengthens your commute. But in late morning, when traffic is light, the highway is effectively nonrival—additional drivers do not affect those already on the road.

Why Do Markets Fail to Provide Public Goods?

These characteristics imply that private individuals and firms, left to their own devices, will undersupply public goods. In other words, the market outcome will fail to be efficient. To see how this happens, let's start with a simple example. Adam and Beth live on either side of a flower garden that they own in common. As far as they are concerned, the garden is essentially a public good, since they both have access to the garden, and one neighbor's enjoyment does not diminish the other's. (Most public goods, of course, involve many more than two people; we're keeping the example simple for the sake of illustration.) Figure 5.3 depicts the marginal cost of tending the garden, along with the two neighbors' marginal benefit curves. The horizontal axis measures the quantity of this public good, which is to say the aesthetic appearance of the garden (its lushness, lack of weeds, plant health, etc.). Note that while both neighbors enjoy the garden, Adam values it more highly than does Beth.

As we did in the cases of the market and of externalities, we start by asking what private provision would yield on its own. In the real world, we might think of this as the "free market" outcome for public goods, in the sense of an unregulated market without government intervention. In our simple example, this corresponds to a lack of cooperation between the neighbors. Left to her own devices, Beth would tend the garden up to the level at which her private marginal benefits equal the marginal cost of provision (denoted Q_B on the graph). At that point, however, Adam's marginal benefit exceeds the marginal cost: Thus he prefers more of the public good, even if he has to bear the entire cost himself. Indeed, Adam will willingly provide the higher amount denoted Q_A. In this case, Beth provides Q_B and Adam supplies the difference $Q_A - Q_B$. Of course, once Beth recognizes that Adam enjoys greater benefits from the garden, Beth will have an incentive not to supply any of the good. Instead, she may choose to free ride on

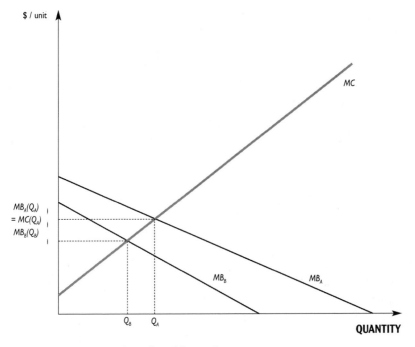

Figure 5.3   Private provision of a public good.

Adam, who would willingly supply the entire amount Q_A on his own. (Note that even when Beth contributes, Adam will not provide more than Q_A, since beyond that point the incremental benefits to him are less than the incremental costs.) As a result, under the free-market outcome Q_A units of the good are produced.

But this presents us with a seeming paradox: Both Adam and Beth would be better off if more of the good were provided! To see why, recall that at Q_A, the marginal cost of increasing the public good just equals the marginal benefit to Adam alone. Thus the *combined* marginal benefit from tending the garden a bit more, to the two neighbors together, must be greater than the marginal cost. Yet as long as Adam and Beth act only in their own selfish interests, without cooperating, neither will make the extra effort: The benefit to either of them alone is too small to make the extra cost worthwhile.

This parable illustrates an important result: Private provision of public goods will be inefficiently low. We have already seen that efficiency requires that marginal benefit equals marginal cost. In the case of the flower garden (or any other public good) the relevant measure of marginal benefit is the social marginal benefit (*SMB*)—in this case, the sum of Adam's and Beth's private marginal benefits. Figure 5.4 compares the efficient outcome with the private provision we found earlier. In the figure, the curve marked *SMB* equals $MB_A + MB_B$. Note that it intersects marginal cost at $Q\star$, a level greater than what Adam provides on his own. Indeed, if Adam and Beth could find an equitable way to share the costs (perhaps if each one knew exactly how much the other valued the garden), and managed the garden jointly, they would tend it up to the efficient level $Q\star$.

The crux of the problem is that the quality of the flower garden is exactly the same for Adam as it is for Beth, regardless of how they divide up the total time spent tending it. The same principle holds for public goods in general. For example, New Haven, Connecticut, has a city park with a new playground and a renovated carousel. The most frequent users, of course, are families with young children; but all city residents enjoy the same free access and hence the same potential consumption. Or consider the example of clean air. The clean air you consume is not measured by the volume of air that passes through your lungs, but rather by the amounts of particulate matter, sulfur dioxide, and so on that are present in the air—concentrations that are the same for all individuals in a given city or region.

Because all individuals experience the same level of the public good, we must sum their marginal benefits when we consider the efficient amount of

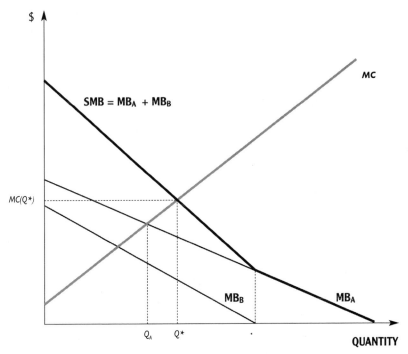

Figure 5.4 Efficient provision of a public good, and underprovision by the market.

the good to provide. This stands in sharp contrast to a market for private goods. Since the benefit from consuming one more unit of a private good will be enjoyed solely and entirely by one person, we need only consider that individual's marginal benefit from the good. Moreover, each person consumes the good until her marginal willingness to pay just equals the price of the good. As a result, every individual in a market for a private good ends up with the same marginal benefit for that good (since all face the same price). In turn, this means that the marginal social benefit of a private good must equal its price, ensuring an efficient outcome.[3]

As a final point, note that public goods problems in the real world are much more complex than our simple two-person example. Neighbors like Adam and Beth might well cooperate; even if not, the difference between the efficient level of the public good and the free market outcome (what Adam provides on his own) may not be all that large. As the number of individuals grows, however, cooperation becomes more and more difficult. Moreover, the marginal benefits of individuals shrink relative to the marginal benefits to society. As a result, the free-riding problem becomes more acute as the size of the relevant public increases, with the gap between the efficient level of a public good and what is privately supplied growing apace.

Public Goods Provision as a Positive Externality

At first glance, public goods may seem to be a fundamentally different problem than externalities. After all, public goods are defined by a pair of specific characteristics, and our discussion to this point has said little about production or cost, which was central to the pollution externality of the previous subsection.

This apparent difference masks a deep underlying connection. In the two-person example above, Beth enjoys the flower garden that Adam alone tends. Adam is not compensated for those benefits, however—that's why we call Beth a free rider, after all. By caring for the garden, therefore, Adam provides a *positive externality* for Beth.

The externality involved in public goods provision plays out on the demand side, while the pollution externality considered before involves the supply side. The market failure from a negative externality such as pollution results from a divergence between private and social marginal costs: Private marginal costs exclude external damages, and thus lie below true social marginal costs. The market failure that arises in public goods provision, on the other hand, comes from the divergence between private and social marginal *benefits*. When the public goods provider (Adam, in our example) considers how much of the public good to supply, he does not take into account the benefits enjoyed by the free rider (Beth), and thus understates the true marginal benefits to society. As a result, too little of the public good is provided. Just as the unregulated market tends to produce too much of a bad thing, so private provision of a public good yields too little of a good thing.

The free-riding problem—in which some individuals under-contribute to a public good, relying on the contributions of others—becomes more acute as the size of the relevant public increases.

The Tragedy of the Commons

Our third category of environmental market failures is known as the "tragedy of the commons," made famous by a well-known article of that name written in 1968 by Garrett

Hardin.[4] A group of people sharing common access to a natural resource will tend to overexploit it, unless they can develop effective government institutions (or social norms) to regulate its use. Hardin used the metaphor of an English pasture, or "commons." The more sheep graze the commons, the less food is available for each of them. Each shepherd bears only a portion of the costs to the commons from the grazing of an additional animal (since those costs, in the form of less fodder, are spread over the herd as a whole)—but he receives the entire gain from increasing his private flock. The result is that each shepherd puts too many sheep to pasture, from the point of view of the commons as a whole.

The tragedy is that the resulting overgrazing reduces the pasture's productivity. As a result, every shepherd would be better off if all could agree to restrict their flocks—but none has an incentive to do so on her own. If one shepherd pares down her flock, another may respond by adding a sheep to his own.

We can delve more deeply into the tragedy of the commons in the context of a very different resource (but one more familiar to modern minds): namely, commuter roads. Consider the commute from a suburb to a hypothetical city. We'll suppose that drivers can reach this city by any one of a number of smaller back roads; this route always takes forty minutes (we'll assume that there are enough of these back roads that they never fill up with traffic).

On the other hand, it takes thirty minutes to reach the city on the highway—at least, if there's no traffic. As more and more drivers use the highway, traffic slows the commuting time. For the sake of this simple example, let's say that each additional driver after the first one slows the commuting time (for every driver) by one second. If 121 drivers use the highway, for example, the commuting time becomes thirty-two minutes (thirty minutes plus 1 second for each of the 120 additional drivers). We'll use this extra commuting time as our measure of the marginal cost of additional drivers. Meanwhile, the marginal benefit from an additional driver is her own time savings from taking the highway rather than the back roads. The net benefit from the highway is then just the total commuting time saved among all drivers, relative to what would happen if they all drove the slower route. (To focus on the problems of congestion and open access, let's set aside the well-known negative externalities associated with automobile emissions.)

What is the efficient number of cars on the highway? Suppose we start with no drivers taking the highway at all. The first driver saves ten minutes in commuting time: that is the marginal benefit. Since there are no other drivers on the road, the marginal cost of the first driver is zero. Now suppose there are already N commuters taking the highway. What happens if we add one more? The marginal cost of the additional driver is one second for everyone else, or N seconds. The highway commute now takes $30 + N/60$ minutes—so the marginal benefit to the $N+1$ driver (the time saved relative to the back roads) is $10 - N/60$.

A little algebra will show you that marginal cost equals marginal benefit when there are 301 drivers. That is, $N^* = 301$ (the asterisk denotes the efficient outcome). At that point, the commute takes thirty-five minutes by highway. The time savings to the 301st driver is five minutes, which is precisely the slowdown she imposes on all the other drivers.

Now what happens if access to the highway is unrestricted? Will the number of cars be efficient? The answer, as you have probably guessed, is "no." To see why, consider what happens in the efficient scenario. When there are already 301 drivers on the highway, every driver taking the back roads thinks to herself: "If I take the highway instead, I will save almost five minutes (actually, four minutes and 59 seconds). Even though I will slow down everyone else, from *my* perspective I will be better off by taking the highway."

What happens if everyone reasons this way? The number of drivers on the highway will continue to increase until there is enough traffic that the highway commute takes exactly as long as the back roads—that is, forty minutes! When each driver rationally makes her decision based only on her own costs and benefits, the aggregate outcome is far from rational for the group as a whole: The net benefits from the highway (the total time saved) ends up being reduced to zero.

The Tragedy of the Commons as a General Model

Hardin's metaphor of the commons applies to many natural resources, but only under two important conditions. First, access to the resource must be unrestricted. Economists describe such resources as *open-access resources*. In terms of our discussion of public goods, an open-access resource can be thought of as a resource that is *nonexcludable* (like pure public goods) but rival. The lack of exclusion usually stems from a combination of institutional and physical factors. A classic example is a deep-sea fishery, such as the cod fishery in Georges Bank in the northern Atlantic, where the distance from shore and the lack of national jurisdiction—along with political obstacles—

make restricting access difficult. Similarly, forest reserves are typically open to harvesting by surrounding populations—especially in the developing world, where the funds necessary to patrol boundaries and prevent poaching are often lacking. Large underground aquifers such as the Ogalalla, which underlies the Great Plains from north Texas to Nebraska, provide a common water source for the farmers and ranchers who live over them. The same can be said of wireless Internet routers. It is easy enough to install security measures, such as requiring a password; but as you no doubt know from experience, open-access networks are common. This is usually not a problem for the owner of the router—unless the residents in an apartment building across the street discover the free Internet access!

A second important condition is *diminishing marginal returns*. In plain English, as the number of people using the resource grows, the benefits from the resource must increase at a slower rate. In our commuting example, the total benefits from the highway increase as more and more drivers use the resource, since they each save time. However, the incremental time savings for each additional driver diminish, because traffic slows down as more and more drivers use the highway.

As in the case of public goods provision, there is a deep connection between the tragedy of the commons and the notion of externalities. In particular, diminishing marginal returns imply that each user of a resource imposes a negative externality on the other users. In the highway example, the negative externality comes in the form of extra commuting time for the other drivers. This "open access externality" implies that unrestricted use will result in the overexploitation of a resource, since individuals will ignore the negative externality their effort imposes on others.

The Tragedy as a Collective Action Problem

The tragedy of the commons represents a particularly stark example of an externality, since when access to a resource is open, overuse tends to drive net benefits to zero. In contrast, of course, net benefits would be positive in the efficient scenario.

In the commuting example, the net benefits at the efficient number of drivers is 301 x 5 minutes = 1505 minutes of commuting time. In the terms of our discussion in chapter 3, a move from open access to the efficient outcome would be a Pareto improvement: Drivers permitted to use the highway would enjoy net benefits from doing so, while those excluded from it would be no worse off than they would be in the open-access case (since the time savings from using the highway are zero when access is unrestricted).

If people vary in their costs and benefits of using a resource, of course, restricting access to it will produce losers as well as winners. This helps explain why governments in the real world have found it so hard to impose restrictions on previously unregulated fisheries, such as the Atlantic cod fishery that once thrived in New England towns. Even so, with appropriate transfers, the increase in net benefits under the efficient scenario means that everyone could be made better off by imposing restrictions on access to the resource. (Recall our discussion of the modified Pareto criterion in chapter 3.)

For this reason, economists (and political scientists) often describe the tragedy of the commons as a *collective action problem*. A collection of individuals—people, or firms, or even nation-states—may find itself in a situation where the group as a whole is better off if all contribute to the common good, but each individual member of the group has incentives to free ride. We can use simple game theory to gain another perspective on the logic behind this dilemma.

Consider the following simple model of two countries considering whether or not to cooperate in cleaning up a pollutant that affects them in common. (For example, think of countries deciding whether or not to sign a binding international agreement that would reduce emissions of carbon dioxide, which contributes to global climate change; for simplicity, we will consider only two countries, but the basic lessons apply to much larger groups.) Let the total cost of effective action be four (the units are merely for illustration). The benefits from such action, however, are three per country, for a total of six. Importantly, if action is taken, both countries benefit, regardless of who paid for the pollution reductions. We assume that the cost of four is split equally if both countries contribute, and is borne entirely by the contributing country if the other refuses.

Cleanup is clearly efficient in this example: total benefits are six, while total cost is only four. However, even though this outcome is jointly optimal, it turns out not to be individually rational. In fact, each country has a strong incentive not to take action. To see why, we need to compare the payoffs to each country under each of the four possible outcomes.

Consider the matrix in figure 5.5. The rows correspond to actions that country A can take—namely, "cooperate" (i.e., help pay for the public good) or "shirk" (pay nothing). Similarly, the columns represent the actions taken by country B. The numbers in each box represent the payoffs to the two

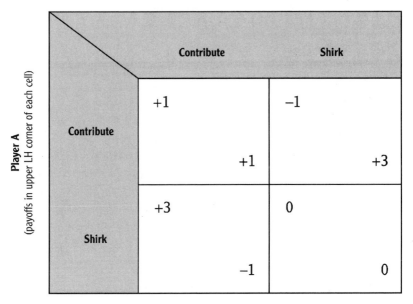

Figure 5.5 Payoff matrix for the Prisoner's Dilemma.

countries, with country A's payoff listed first. For example, if both countries agree to contribute, they each receive a payoff of plus one: a benefit of three minus their share of the total cost, which is two. If both shirk, they each receive zero. Clearly, each country is better off if both contribute than if neither does.

But the relevant question facing each side is: How does *my* action affect *my* payoff? Suppose you are country A, choosing between contributing and shirking. Suppose that you expect country B to contribute. In that case, you would get a payoff of plus one from contributing, and would receive plus three from shirking; thus shirking yields a higher payoff, and is individually rational. What if you expect B to shirk? In that case, contributing yields minus one (benefits of three minus a cost of four), while shirking yields

Economists (and political scientists) often describe the tragedy of the commons as a collective action problem.

zero; so shirking is again preferable to contributing! The same logic holds for country B. Thus regardless of what action the other country takes, each country does better by shirking. In the language of game theory, shirking is a "dominant strategy."

The game we have described is an example of a "Prisoner's Dilemma," commonly used to describe collective action problems—not just in the environmental arena, but in a broad range of social and economic situations. (The name evokes a district attorney who induces two criminals in separate cells to rat out each other, by promising each that he will get off easy if he confesses to the crime while the other stays silent.) Individually rational decisions produce a collectively suboptimal outcome. In the case of an open-access fishery, for example, every individual fisher has an incentive to keep fishing as long as her private gains from doing so outweigh the costs. But when all fishers reason in this way, the net benefits of the resource are driven to zero. As a result, they would all be better off if they could restrict access to the efficient number.[5]

Conclusion

If chapter 4 extolled the potential virtues of markets, our extensive discussion of market failures in this chapter is an important reminder that laissez-faire markets are a prescription for environmental problems. When markets are incomplete, individuals face the wrong incentives to change their behavior—to reduce the pollution they produce, or contribute to the provision of public goods, or refrain from exploiting a common resource.

While we have discussed negative externalities, public goods, and the "tragedy of the commons" in turn, it bears emphasis that these are not distinct problems, but rather different ways of framing the same underlying market failure. The key to all three is the notion of *nonexcludability*. An externality arises when the good or ill effects of one person's actions, or a firm's operations, are not borne exclusively by that person or firm. People who cannot be excluded from enjoying a public good may prefer to free ride on what others provide rather than contributing to the good themselves. And those with open access to a common resource have little incentive to moderate their use of the resource, knowing that others will take what they do not.

Viewed in this way, economics provides a morally neutral explanation for environmental problems. Pollution and overfishing do not arise because polluters or fishers are bad people. Rather, the managers of polluting firms, like fishers, are simply trying to maximize their profits. (Indeed, the managers of private firms have legal responsibilities to their shareholders to maximize the firm's profits.) We might see this as selfish behavior, as indeed it is; but we have seen that selfish behavior by itself is not the problem, since in well-functioning markets it is the engine of efficiency. To an economist, the root cause of environmental problems concerns the *incentives* people face. The driving factor is not that individuals pursue their own interests, but rather that in an unregulated market nothing aligns self-interest with the broader effects on society.

Framing the problem as a problem of incentives—rather than as a problem of morality, or of markets per se—also points the way toward possible solutions. As we will see in chapter 8, economic theory suggests that the way to deal with market failures in the environmental realm is not to avoid the use of markets, but rather to fill in the incomplete nature of the market, whether by providing artificial price signals, assigning property rights, or even creating a market in environmental goods such as clean air.

6

Managing Stocks: Natural Resources as Capital Assets

Many people assume that natural resources have infinite value. But economics does not, as we discussed in chapter 2. As a general matter, economists treat natural resources as a subset of society's capital assets, no more or less important than other types of capital. This human-centered approach is fundamentally different from other perspectives, like deep ecology and intrinsic rights. You may or may not find it difficult to reconcile this approach with your own values with respect to natural capital. Nonetheless, we will approach natural resource management problems from the perspective of economic efficiency. This approach highlights the trade-offs that must be made among competing uses of scarce natural resources—for example, recreation, species habitat, and timber extraction, in the case of a forest.

In this chapter, we discuss the economics of nonrenewable natural resources such as oil and minerals. We begin with a discussion of scarcity as an economic concept, which incorporates more than simply the limited availability of physical resource stocks. We then present a simple two-period model of nonrenewable resource extraction, using it to understand the economic notion of scarcity and how resource owners will take it into account as they decide how much to extract. We develop the concept of marginal user cost, an extra cost of extracting nonrenewable resources that represents the opportunity cost of foregone future consumption. Finally, we stress the critical role of property rights in determining whether nonrenewable resource extraction in real-world markets will be efficient.

Economic Scarcity

Natural resource scarcity has economic and geologic dimensions. The critical point in the economic analysis of resource management is this: the "stock" of a natural resource, like oil, depends not only on the physical availability of that resource within the earth's crust, but also on its marginal extraction cost and the prices people are willing to pay to purchase it. For example, some oil fields in Texas are productive only when energy prices are very high. At other times, these fields lie idle because they cannot profitably be operated. The tar sands of Alberta, where oil can be produced by sucking sticky tar out of sandy soil, are viable commercial sources of oil only when the market price of crude oil rises above thirty dollars per barrel. Effective stocks of natural resources continually expand and contract in response to technological change and resource prices, with high prices increasing the quantity of resources that are worth extracting, and low prices reducing it.

A useful way of representing both the economic and physical dimensions of resource stocks, known as a "McKelvey diagram," is presented in figure 6.1. The original McKelvey diagram was developed by the U.S. Geological Survey to classify the stock of nonrenewable resources along its physical (horizontal) and economic (vertical) dimensions. Earth's total resource endowment is, of course, both unknown and fixed—it has only physical dimensions. But the portion of this endowment that is potentially useful to humans depends on both geological availability and economic value. In addition, technological progress continually affects both the costs of resource extraction (generally driving down these costs) and the value of specific resources to society. For example, mechanization and large-scale surface mining have lowered the cost of coal extraction over time. But the advent of gasoline-powered engines made coal obsolete in certain uses.

In thinking about natural resources from an economic perspective, we must be careful in interpreting some common indicators of physical scarcity, such as the reserve-to-use ratio or static reserve index. These ratios divide current known reserves of a resource by current annual consumption, thus measuring the number of years until the resource is exhausted. The problem is that such measures ignore the economic dimensions of scarcity, and so offer an inaccurate picture of resource limits. For example, despite substantial increases in annual consumption, static reserve indices for iron,

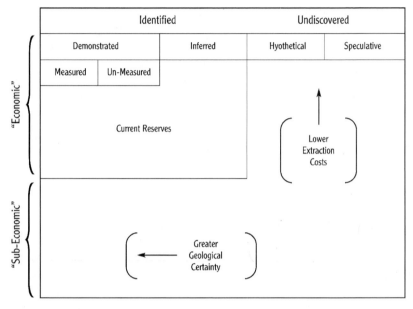

Figure 6.1   McKelvey diagram.

copper, aluminum, and zinc all increased over the period between 1950 and 2000. These increases may reflect exploration for and discovery of new reserves; technological change, which can lower extraction costs, making known reserves exploitable where previously they were not; and increasing commodity prices.

Failing to take the economic dimensions of scarcity into account is a common mistake—so common that even some prominent economists throughout history have made it. Stanley Jevons, a renowned nineteenth century British economist, predicted in a book called *The Coal Question* that as Britain depleted its coal reserves, its economic power would decline precipitously. Jevons failed to account for the fact that an increase in the price of coal would spur the development of alternative sources, new extraction technologies, and greater efficiency in coal use, and he failed to anticipate the rise of oil and natural gas as alternative fuels.[1]

While physical scarcity is only one dimension of economic scarcity, the limited physical availability of nonrenewable natural resources does affect their optimal rate of use. In particular, faced with limited stocks, maximizing the net benefits of a resource to society will gradually require using less of it today and keeping more in the ground to use in the future than if the resource were in infinite supply. In addition, the market prices of nonrenewable resources will be higher than they would be if stocks were not limited, reflecting the impact of scarcity. Let's use a simple example of an oil well to look at the use of a resource over time and to demonstrate these dual effects of scarcity on the efficiency problem.

Efficient Extraction in Two Periods

Suppose we own an oil well, and we plan to pump oil from the well in two time periods—"today" and "tomorrow."[2] (The labels "today" and "tomorrow" are just for convenience; the goal of the model is to explore the balance between current and future use of a scare resource.) The demand for oil in each period is $MB = 10 - .5q$, where q is the quantity extracted; the marginal cost of extracting a barrel of oil (which might include labor and electricity, for example) is constant at $MC = \$3$.

First, let us assume that our oil supply is not limited, but infinite. What would be the efficient quantity of oil to extract today? In order to figure this out, we would set the marginal benefits of extracting oil today equal to the marginal costs.

$$MB = MC$$
$$10-0.5q = \$3$$
$$q^* = 14 \text{ barrels}$$

Using this static efficiency rule as developed in chapter 2, we would extract fourteen barrels of oil today.

Now, let's introduce a limited stock—only twenty barrels are available. If we extract fourteen barrels today, as we would like to, what would that leave for tomorrow? We would be left with only six barrels of oil in the ground. Given no change in demand and marginal cost between today and tomorrow, if we apply the static efficiency rule again tomorrow, we will want to pump another fourteen barrels. But our remaining six barrels will fall well short of this goal (see figure 6.2).

Has the efficiency rule failed us? The problem we have just solved twice sequentially is *myopic*. We have intentionally ignored the limited oil supply and acted as though extraction of oil today is independent of the quantity

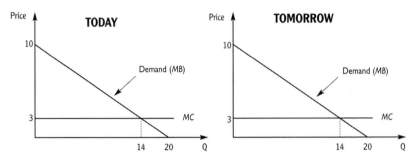

Demand: MB = 10-0.5q

Figure 6.2 The problem with (myopic) static efficiency in the case of a scarce resource. Static efficiency would imply extracting fourteen barrels in each period—more than the total stock of twenty barrels.

left to extract tomorrow. In doing so, we have not identified a loophole in the efficiency rule. We have simply left out a very important cost from our scenario.

When a resource is finite, like our oil well with a mere twenty barrels, one cost of extracting a unit of that resource is the lost opportunity to extract that unit in the future. In addition to the marginal cost of extracting a barrel of oil, we must account for the marginal cost of using up a barrel of oil, which leaves one fewer to use in the future. This extra cost is called *marginal user cost*, or *scarcity rent*. Accounting for the marginal user cost associated with oil extraction, like any cost increase, will reduce the amount of oil that we can efficiently extract today, leaving more in the ground for tomorrow.

Let us solve this problem again, this time taking the limited stock directly into account. The dynamic two-period problem we now solve differs from the static efficiency problem above in three important respects. First, because we are interested, today, in the value of extracting oil both today and tomorrow, we will need to discount the returns to oil extraction tomorrow to reflect the time value of money. This will help us to account for the fact that any oil left in the ground until tomorrow cannot be sold on the market today, and the proceeds from its sale cannot be invested to increase in value between the two periods. Thus the marginal benefits and marginal costs of oil extraction will be expressed in terms of *present value*—their value in today's dollars.

Second, we will introduce the stock constraint directly into our efficiency problem. To do this, we will define the quantity of oil available to extract tomorrow, q_2, as the difference between the total stock (twenty barrels) and the amount extracted today, q_1.

Third, rather than setting the marginal benefits and marginal costs of extraction in a single period equal to each other, we will equate the net marginal benefits (benefits, less costs) of oil extraction in each period. That is, we will start from the presumption that, in order to maximize the net benefits of this oil well, we must ensure that the net benefit of the last barrel pumped today is equal to the net benefit of the last barrel pumped tomorrow. If this were not the case, we could increase the overall net benefit of the oil well by redistributing our pumping plans over time. We will explore this assumption in detail later in the discussion. Below, we solve for the efficient quantities of oil to extract today and tomorrow, assuming a discount rate of 10 percent.

$$PV(MB_1 - MC_1) = PV(MB_2 - MC_2)$$

$$10 - 0.5q_1 - 3 = \frac{10 - 0.5q_2 - 3}{1 + .10}$$

$$7 - 0.5q_1 = \frac{7 - 0.5(20 - q_1)}{1.10}$$

$$7 - 0.5q_1 = \frac{0.5q_1 - 3}{1.10}$$

$$q_1{}^\star = 10.19 \; barrels$$
$$q_2{}^\star = 20 - q_1{}^\star = 9.81 \; barrels$$

We suggested earlier that the efficiency rule in the presence of resource scarcity would cause us to use less of a resource today than we would if the resource were infinite. That is certainly the case in our oil well example. When we solved this problem myopically, thinking only about the net benefits of extracting this oil today, heedless of dwindling stocks, we planned to extract fourteen barrels of oil. Now that we have incorporated the limited stock into our problem, the rules of efficiency tell us to extract just over ten barrels of oil today, leaving the rest in the ground for tomorrow.[3]

A Closer Look at the Efficient Extraction Path

Why not split the well's contents exactly in half, extracting ten barrels today and ten tomorrow? The time value of money is the reason we extract just over half today and just under half tomorrow. Because the value of the oil we extract today can earn interest in an alternative investment between today and tomorrow, it is efficient to extract a bit extra today. In fact, if you experiment with interest rates other than the 10 percent assumed here, you will notice that the higher the interest rate, the greater the difference between the amount of oil we will extract today in an efficient scenario, and the amount that we will leave for tomorrow. In seeking to maximize the net benefits of this oil well to society, these two different types of capital—oil and money—are fungible.

In fact, there are an infinite number of extraction quantities, today and tomorrow, that sum to our twenty-barrel limit. Why is the specific extraction path we arrived at—10.19 today and 9.81 tomorrow—the efficient one? A diagram may help illustrate the intuition behind these numbers. Figure 6.3 plots the marginal net benefits, in present value terms, of oil extraction in each period in figure 6.3. Oil extraction today increases along the horizontal axis from left to right, and extraction tomorrow increases from right to left. The two marginal net benefit curves intersect at the efficient allocation of extraction over time—the pair of extraction quantities for which we have just solved algebraically. Since the curves on this graph represent marginal net benefits, the total net benefits of this resource to society are measured by the area under these curves. And we simply cannot generate greater total net benefits by choosing any extraction path other than the efficient path we have identified. If we move to the right of the efficient allocation, extracting more today and leaving less for tomorrow, the value of net benefits lost tomorrow would exceed today's gains. And if we move to the left of the efficient allocation, extracting less today and leaving more for tomorrow, the value of net benefits lost today would exceed those gained tomorrow.

Marginal User Cost, a "Special" Externality

Earlier, we mentioned that extraction of scarce resources, like oil in a finite well, imposes a cost above and beyond the marginal cost of extraction—a marginal user cost. Is it possible to identify this extra cost, either in our algebra problem or in the diagram? The marginal user cost of a barrel of oil

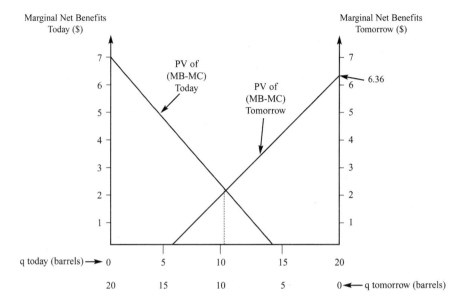

Figure 6.3 Nonrenewable resource extraction: the two-period model.

from our well becomes apparent when we solve for the prices that we can expect to collect for a barrel of oil in each period, today and tomorrow:

$$p_1{}^\star = 10-0.5(q_1{}^\star) \approx \$4.905$$
$$p_2{}^\star = 10-0.5(q_2{}^\star) \approx \$5.095$$

The market price of a barrel of oil is approximately $4.91 today and $5.10 tomorrow, yet the marginal extraction cost is only $3. If you have taken an introductory microeconomics course, it may appear as though we have just violated one of the fundamental tenets of a competitive market— that the price received for a good or service should be exactly equal to the cost of producing the last unit, or price equals marginal cost. This difference between price and marginal cost in the case of scarce resources like oil is, in fact, the user cost we discussed earlier.

When resources are limited, current consumption comes at the cost of foregone potential future consumption. The present value (at the margin) of these foregone future consumption opportunities is marginal user cost, or scarcity rent.

When resources are limited, current consumption comes at the cost of foregone potential future consumption. The present value (at the margin) of these foregone future consumption opportunities is marginal user cost, or scarcity rent.

We can also think of marginal user cost as a negative externality to current oil consumption. Extracting today, we impose an extra cost on tomorrow—diminished supplies. This is not true only of oil, of course. If residents in Las Vegas, Nevada, an extremely arid city in the western United States, use large quantities of water to grow lush, green lawns, this may involve no scarcity rent if the water is from a large, quickly replenishable supply. However, if lawn watering draws down nonrenewable groundwater supplies, then this extra cost of diminished future supplies (lower aquifer levels) should be incorporated into water prices.

Water prices are rarely determined in a market. But oil prices usually are. In our example, the marginal extraction cost does not change between today and tomorrow, but the market price of a barrel of oil rises. Thus it appears that user cost (the difference between price and marginal cost) rises over time. This fact helps us to understand the alternate name for user cost—scarcity rent. As our stock of oil dwindles, oil becomes scarcer, thus scarcity rent—the extra cost of using up a barrel of oil today, which the owner collects in the form of a higher price—increases.

Earlier, we used a McKelvey diagram to demonstrate the physical and economic dimensions of scarcity. Marginal user cost is an economic indicator of scarcity that takes into account both the known physical limits of a resource and what we are willing to pay for that resource. If marginal user cost is an economic indicator of scarcity, how can we be sure that it will really be incorporated into market prices? After all, we have discussed many examples of externalities in this text, and in most of those examples, markets fail to account for environmental and resource damages, leading to inefficient outcomes and often requiring government intervention.

The answer to this question depends critically on the structure of property rights with respect to a scarce resource. Note that, if we own the oil well we have been discussing, by extracting oil, we impose a marginal user cost on *ourselves*, diminishing our own future supplies. Thus we have a strong incentive to account for that cost as we decide how much oil to extract! If we do not, we will not maximize the profits from our oil resource over time, and in a competitive market we will soon be out of business. This is a strong contrast to the examples that we discussed in chapter 5, in which environmental costs were borne by parties other than the externality-generating firms, themselves.

So when nonrenewable resources are privately owned and extracted in a competitive market, resource owners will account for scarcity in determining the optimal timing and quantity of extraction (the extraction "path"). They will treat oil resources, and other nonrenewable resources, like any other capital asset in their portfolio—as stocks that generate returns by the very nature of their scarcity.

The Hotelling Rule

In fact, when we consider nonrenewable resources as capital assets, it is clear that they must generate these returns at a very specific rate. Nonrenewable resource stocks should increase in value at a rate equal to that of other types of assets in the market. A useful benchmark here is the prevailing rate of interest, which represents the risk-free return an investor can earn in the market. If oil stocks in the ground were gaining in value at a rate faster than the rate of interest, resource owners would extract nothing in the near term, leaving stocks in the ground to increase in value relative to money in the bank. If oil stocks in the ground were gaining in value at a rate slower than the rate of interest, resource owners would do better to extract all of the oil, sell it, and invest the proceeds.

Let us test this theory with our oil well example. We can calculate the rate of change in marginal user cost between the two periods as follows:

$$\frac{MUC_2 - MUC_1}{MUC_1} = \frac{(P_2 - MC_2) - (P_1 - MC_1)}{(P_1 - MC_1)}$$

$$= \frac{(5.095 - 3) - (4.905 - 3)}{(4.905 - 3)}$$

$$\approx 0.10$$

This interesting theoretical result, that marginal user cost rises at the rate of interest, is called the Hotelling Rule, named for statistician and economist Harold Hotelling. The key to understanding the Hotelling Rule lies in realizing that in a competitive market the limited availability of nonre-

This interesting theoretical result, that marginal user cost rises at the rate of interest, is called the Hotelling Rule.

newable resources like oil strongly affects resource prices and extraction paths. Oil in the ground generates returns for its owner over time. It is a capital asset that can be spent today (through extraction) or saved for tomorrow. The price of spending it today is the lost scarcity rent it would generate by remaining in the ground. Likewise, the return to saving it for tomorrow is the rate of increase in scarcity rent.

If a market is in dynamic equilibrium, private owners of capital cannot increase their profits by reallocating their portfolios—if they could make more money by holding less capital in oil and more in some other asset, or vice versa, private owners would take advantage of that opportunity. The intuition here is like a "no-arbitrage condition." The competitive pressures of the market for privately owned nonrenewable natural resources are extremely powerful, and thus one particular negative externality to depleting these resources—the fact that they will not be around to consume tomorrow—will be reflected in market prices and extraction rates.[4] In this sense, at least, the markets for nonrenewable resources are complete.

Of course, there are other externalities generated through the extraction and consumption of petroleum—this should be clear from our earlier discussion of the economics of pollution control. And we have not reached the bottom line in our discussion of the impact of scarcity on markets; we will return to the potential impact of dwindling resource stocks on human welfare when we discuss the economics of sustainability in chapter 11. But the Hotelling Rule helps us to understand why economists tend not to worry about the extraction decisions of private owners of nonrenewable resources in competitive markets.

The Critical Role of Property Rights

What gives us this confidence that the individual decisions of private owners of nonrenewable resources will maximize the present value of nonrenewable resource stocks to society? The answer lies in the structure of property rights with respect to this class of resources. With few exceptions, nonrenewable resources like oil and other minerals tend to be privately owned and traded in reasonably competitive markets. Recall that marginal user cost is basically an externality that is really not external to the transac-

Oil Extraction in Noncompetitive Markets

One of the key players in the global oil market is the Organization of Petroleum Exporting Countries (OPEC), a cartel that calls into question our assumption of perfect competition. How does market power, in the form of oligopoly or monopoly, affect our basic result? In general, monopolists and cartels (when they are successful) increase their profits above that which they would achieve in competitive markets by restricting output to raise market prices. They do this, in theory, just to the point at which the extra returns from higher prices equal the lost returns from lower output. The situation is no different in the market for oil—individual production quotas for member countries are, in fact, OPEC's chief tool for earning extra profits. When this strategy succeeds, oil extraction should be slower than the dynamically efficient extraction rate identified by the Hotelling Rule, and prices higher. Thus, Nobel laureate Robert Solow has dubbed the monopolist the "conservationist's friend," if a conservationist is one who supports the use of natural resources at slower-than-efficient rates. As is typically the case when one of the major assumptions required for market efficiency fails to hold, the net benefits to society of an oil resource in the presence of market power are not maximized.

tions between buyers and sellers of oil—sellers incur this consumption externality (reduced future supplies) themselves, and thus will take it into account in determining how much oil to extract.

Were we to have used groundwater aquifers, rather than oil wells, as our example in the preceding discussion of nonrenewable resources, we would not have arrived so cleanly at the Hotelling result. We can make this general statement because, in contrast to mineral resources, groundwater supplies tend not to be privately owned and traded in competitive markets. When a farmer pumps water from a nonrenewable groundwater aquifer, the marginal user cost associated with pumping a unit of water is not incurred by that farmer; instead, the cost of diminished future supplies is spread among all of those who benefit from the aquifer. Thus, individual users of the resource have no incentive to take scarcity into account when deciding how much water to pump.

What would happen if a forward-thinking farmer did try to save a cubic meter of water in the aquifer for tomorrow, rather than pump it today? Unlike the owner of the oil well, the farmer cannot assume that this unit of water will remain in the ground for his own benefit tomorrow—it is much more likely that another user of the resource (perhaps one of his competitors down the road) will pump that unit of water, spoiling our farmer's good intentions. It is easy to see how we might arrive at a race to pump, quickly draining the resource, rather than extracting it in a dynamically efficient manner. Economists refer to this type of good—one from which potential consumers cannot be excluded, and one that consumers compete to capture—as an *open-access* resource, as discussed in chapter 5.

While a nonrenewable groundwater aquifer would seem to be directly comparable to a nonrenewable oil well, the difference in property rights regimes between the two cases leads to very different outcomes. In the case

An Open-Access Resource: The Ogallala Aquifer

A good example of an open-access resource is the Ogallala Aquifer, which underlies approximately 174,000 square miles in the U.S. Great Plains, including portions of the states of Texas, New Mexico, Kansas, Nebraska, Colorado, Oklahoma, South Dakota, and Wyoming. The aquifer contains approximately 3.8 billion acre-feet of water and provides about 30 percent of all groundwater used for irrigated agriculture in the United States. One economic analysis, based on the variation in land values between irrigated and dryland farms in the region, estimated that the water's value ranges from 30 to 60 percent of the irrigated farmland sale price in the region.[5]

In most states that sit on top of the aquifer, groundwater is an open access resource. Unlimited quantities of water can be extracted by individual farmers, who incur only the costs of extraction (the water itself is free). Use of the aquifer for irrigation has proceeded at a rate conservatively estimated to be about ten times the rate of recharge (which in many parts of the aquifer is negligible). Since the resource was first exploited on a large scale in the 1940s, groundwater levels have dropped precipitously in some areas, particularly north Texas, Oklahoma, and southwest Kansas.

Because the Ogallala is an open-access resource, the marginal user cost associated with pumping water from the aquifer is truly an externality—no party incurs the full cost of pumping, and thus the resource is depleted at a rate faster than the dynamically efficient rate. Collectively, society would be better off if irrigators did take into account marginal user cost, the marginal cost of aquifer depletion. But individual irrigators have no incentive to internalize this cost when making their water use decisions. Thus, the Ogallala aquifer will be depleted inefficiently soon.

of oil, we expect markets to do a reasonable job ensuring dynamically efficient extraction; in the case of groundwater, we expect markets in their current structure to fail at this task.

Conclusion

In this chapter, we have introduced the first application of dynamic efficiency in the determination of optimal extraction rates for nonrenewable natural resources such as oil and coal. While such resources are available in limited quantities on the earth, the economic concept of scarcity that we developed in this chapter takes many things other than physical limits into account, most importantly the effects of rising prices on demand for scarce resources.

We showed how private, competitive owners treat nonrenewable resource stocks as capital assets. In doing so, they extract resources at a rate that takes into account the limited physical stocks. The Hotelling Rule told us that this optimal extraction rate of nonrenewable resources maintains an asset market equilibrium, in which the rate of return to stocks in the ground equals the rate of return to alternative investments.

In this first natural resource management application, we encounter a situation in which real-world market outcomes do a pretty good job at approximating efficient outcomes, in large part due to the fact that nonrenewable resources tend to be privately owned and traded in reasonably competitive markets. We ended with a reminder that, where property rights are not well-defined, extraction rates of nonrenewable resources can be expected to exceed efficient extraction rates. The situation of poorly-defined property rights will be even more relevant to the discussion of renewable resources in chapter 7.

7

Stocks that Grow: The Economics of Renewable Resource Management

Many renewable resources are more like an aquifer than an oil field. That is, they often are open-access or common property resources, and we do not expect them to be used in a dynamically efficient manner when markets are left to their own devices. This absence of property rights complicates economic analysis. A second complication is that the stock of a renewable resource, while limited, is not fixed. For example, the stock of fish in a fishery might be a certain number at any particular moment, but over time it depends on factors like reproduction rates and predation, including human fishing effort.

In the case of nonrenewable resources like petroleum, we were concerned with calculating the dynamically efficient rate of depletion of the resource. In contrast, with renewable resources we hope to calculate the size and/or timing of the efficient harvest, in many cases maintaining a sustainable flow of the resource in perpetuity. Because renewable resource stocks are functions of both natural systems and human behavior, the models that we use to analyze them combine biology and economics—they are *bioeconomic* models. The first such model that we will discuss is a bioeconomic model of a forest.

Economics of Forest Resources

When we were considering how fast to pump oil in the preceding chapter, we concluded that efficient management of an oil well (or any nonrenewable resource) requires extraction at a rate that maximizes net benefits. Similar logic applies to forests. In economic terms, standing trees are capital assets that increase in value as they increase in volume over time. But allowing the trees to stand is also costly—we must consider the opportunity cost of alternative investments. Thus, we seek to identify the length of time to wait between timber harvests that maximizes the difference between total benefits and total costs (in present value).

The economic analysis of forest management raises two issues that were largely absent from our discussion of nonrenewable resources. First, oil and coal have value primarily as inputs to the production and consumption of other goods, like energy. In contrast, the value of a forest is more complex. In addition to their value as timber for potential harvest, standing trees offer other benefits, providing species habitat and carbon sink. To keep things simple, we will start with the problem of commercial timber extraction, and we will add other types of values later.

Second, forested lands exhibit a wide variety of property rights regimes, ranging from private ownership to open access. We start by considering a private landowner who makes rent-maximizing decisions about harvesting her trees. Later in the chapter, we will consider other property rights arrangements.

Forest Growth and the Biological Rotation

We begin with a simple model of forest growth. We can model the volume of timber in a stand of homogeneous trees as a function of time. Here we use the following volume function, pictured in figure 7.1, to describe this process:

$$V(t)=10t+t^2-0.01t^3$$

At first, the rate of growth is very fast. Over time the trees continue to grow, but the rate of growth begins to decline (in our model, after about thirty-three years). At some point, depending on the species, climate, and a variety of other factors, the trees stop growing and begin to decay, resulting in declining volume (in our model, after about seventy-one years).

One candidate for the best interval at which to cut and replant these trees is the age that maximizes the *mean annual increment* (MAI), the average volume of the stand, $V(t)/t$. If we divide volume by time, we obtain the MAI curve depicted in figure 7.2, which reaches its maximum after fifty years of stand growth. This decision criterion makes some intuitive sense, because no other rotation yields a greater average volume of wood. For this reason, the maximum MAI is often called the *biological rotation*.

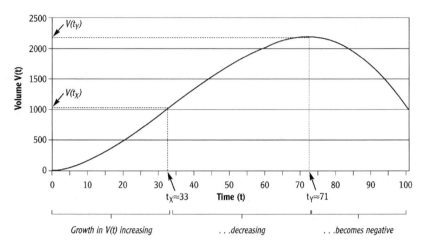

$$V(t)=10t+t^2-.01t^3$$

Figure 7.1   Timber volume in a forest as a function of time.

Does the biological rotation maximize the net benefits of the stand to society? We haven't yet introduced any economic information into our discussion, so you should guess that the answer to that question is "probably not." Figuring out the efficient rotation requires that we think about trade-offs, as we did in the case of petroleum. For example, if we were to cut the trees after forty years, instead of fifty, we would obtain fewer board feet of timber, but we would obtain the smaller cut ten years sooner. Given the time value of money, this might make sense. Now we introduce some economic information in the simplest possible case—a single harvest.

Optimal Aging Problem: The Wicksell Rotation

Say that we are interested in the returns to harvesting this stand of trees once, with no concern for what will happen to this currently forested land after we extract our timber. This single rotation problem is essentially an "optimal aging" problem. The question "How long should I age a stand of hardwoods?" is quite similar to the same question involving a bottle of wine, or a fine cheese. To answer it, we must think about the returns to alternative investments, again represented by the rate of interest.

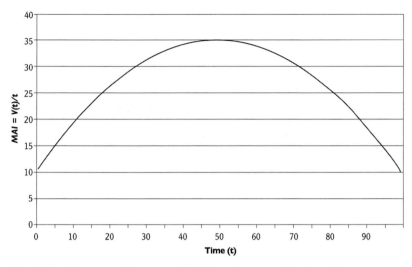

Figure 7.2   Mean annual increment $(V(t)/t)$ in a forest, as a function of time.

To solve this problem, think of the situation a private landowner would face each year. She would compare the net returns to cutting her trees this year to the net returns to waiting for one more year. As long as the net returns to cutting now were less than the net returns to waiting, she would prefer to keep her assets in standing trees. The net benefit-maximizing year in which to cut the trees would occur just as the net returns to waiting equaled the net returns to cutting. We can represent this point in a simple equality, in which the net returns to cutting now are on the left-hand side, and the net returns to waiting (in present value) are on the right:

$$(p-c)\,V(T_0) = \frac{(p-c)\,V(T_1)}{(1+r)}$$

where:

p = timber price
c = unit harvesting cost
$V(T_0)$ = stand volume this year
$V(T_1)$ = stand volume next year
r = discount rate

If we rearrange some terms, we obtain the following:

$$r = \frac{V(T_1)-V(T_0)}{V(T_0)}, \text{ or } r = \frac{\Delta V}{V(T_0)}$$

Thus, it is efficient to harvest the stand when the rate of growth in timber volume, the rate of return to our capital asset (standing trees), is equal to the interest rate.

This is called the Wicksell Rule, and it can be applied to any optimal aging problem. If we harvest the stand before this point, the lost value of the incremental growth we would expect between this year and next would exceed the value of the incremental gains we would earn by depositing our net harvest proceeds in the bank to earn interest for one year. If we wait to harvest the stand beyond this point, the opposite would be true. The Wicksell Rule, like the Hotelling Rule, is a no-arbitrage condition. Just as in the case of oil extraction, if forest owners could make more money by holding less capital in trees and more in some other asset, or vice versa, they would take advantage of that opportunity.

There is an inverse relationship between the Wicksell rotation and the rate of interest. If the expected returns to alternative investments are very low, the Wicksell rotation is very long; a high interest rate implies a shorter rotation. For interest rates above 2 percent, the Wicksell rotation for a stand of trees described by the volume function in figure 7.1 is shorter than the biological (MAI-maximizing) rotation. Note that incorporating the time value of money into our model had the predicted result. As we guessed at the end of the previous section, under reasonable assumptions about the interest rate, we would prefer to accept a smaller total volume than that afforded by the biological rotation in exchange for cashing in our trees at an earlier date.

Efficient Forest Management over Time: The Faustmann Rotation

The time value of money is not the final wrinkle in the problem of optimal forest rotation. We have one more concern, which we did not worry about when discussing oil extraction—the value of the land on which our trees are growing. The problem of optimal rotation is really one of optimal land use. A landowner deciding when to harvest a stand of trees is concerned not only with the growth rate in the value of alternative assets—that is, how much she might earn by cashing in her trees once and putting the money in the bank—but also with the value of her property as a whole. The problem requires an understanding of ongoing returns to forestry on a tract of land over time, and a comparison of these returns to those from other potential land uses.

> *The Wicksell rule for the optimal single rotation tells us that it is efficient to harvest a stand of trees when the rate of growth in timber volume, the rate of return to standing trees, is equal to the interest rate.*

The landowner who is mindful of the value of her resource as a whole faces a variety of choices each year. She could cut her timber and replant; she could wait one more year, then cut and replant; she could cut this year and convert the land to a new use, like planting watermelons or building suburban tract housing; or she could cut this year and sell the land to a new owner.

To solve the optimal rotation problem, which takes all of these options into account, we introduce the concept of *site value*.[1] Site value is the value of a forested piece of land, assuming that the landowner will implement efficient forest rotation in perpetuity; or—if forestry is not the most profitable use of that land at any point in the future—convert the land to its most profitable use. Site value allows us to compare the present value of expected future rents (benefits less costs) from forestry to those from other potential land uses, like farming or residential development. In economic analyses of land use and land-use change, this is how land prices are determined. Land prices are equal to the present value of expected future rents from land in its most profitable use. Thus, site value, which we will represent here using the letter S, captures all of the competing land-use options we mention above.

When the landowner considers site value, as well as the annual return from cutting and selling her timber, her yearly problem looks a bit different than the one we defined above. As before, she will seek to cut her timber and replant in the year in which the marginal net benefits of cutting are equal to the marginal net benefits of waiting one more year. We can represent this point in a simple equality, in which the net returns to cutting now are on the left-hand side, and the net returns to waiting one more period (in present value) are on the right. Everything is as before, but we have added two additional terms, site value (S) and the cost of replanting trees after the timber harvest (D).

$$(p-c)V(T_0)-D+S = \frac{(p-c)[V(T_0)+\Delta V]-D+S}{(1+r)}$$

The easiest way to understand the intuition behind including site value on both sides of this equation is to think of S simply as the sale price of the land. The net returns to cutting in each period include not only the per-unit returns from timber less the cost of replanting, but the amount of money the landowner would make if she sold her land immediately after replanting. Even if the landowner has no plans to sell this land, S still must be included in each year's potential returns, because it represents the opportunity cost, to her, of holding this land in forest, rather than doing something else with it.

To simplify things, let us now consider $V(T)$ to represent not simply the volume of timber in the forest at time T, but the net value of that volume, or $(p-c)V(T)$. In this case, we can reduce the equality above to:

$$r[V(T_0)-D]+rS = \Delta V$$

The left side of the equation is the marginal benefit of harvesting now. The right side is the marginal cost of harvesting now (the extra timber volume that would accrue between this year and next, which the landowner forgoes through her impatience). Figure 7.3 illustrates efficient timber rotation. In early years, timber volume in the forest is growing quickly, so the increase in value is relatively large; as growth slows, the benefits of cutting the forest approach the costs. At $T^\star$ the marginal benefits and costs of cutting are exactly equal. If the landowner rotates this stand at $T^\star$ in perpetuity, she will maximize the net benefits of the forest resource.

To continue with our discussion of efficient natural resource management guidelines as "no-arbitrage conditions," we can rearrange terms to obtain:

$$r = \frac{\Delta V}{V(T_0)-D+S}$$

The landowner should time her harvests so that the rate of return to her forest assets is equal to the prevailing rate of interest. Notice that, in contrast to the Wicksell Rule, here the landowner is interested in the rate of return to the value of her forested land, not just the timber volume. Thus we

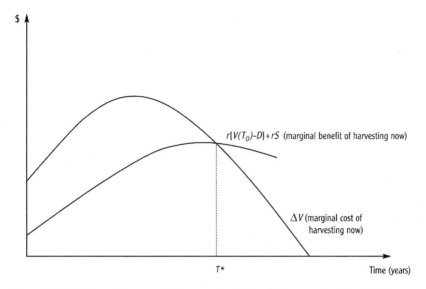

Figure 7.3 Efficient (Faustmann) forest rotation. The efficient rotation length, T^*, equates the marginal benefit of harvesting and the marginal cost.

include S in the denominator. This rule of thumb for efficient forest rotation, taking into account both the time value of money and the opportunity cost of land, is called the Faustmann Rule.

We noted earlier that the biological rotation will, in general, be longer than the Wicksell rotation. How do these compare to the Faustmann rotation? In general, the Faustmann rotation will be the shortest of the three. Why is this the case? This should be obvious from a comparison of the Wicksell and Faustmann rules, given that the Faustmann rule has a larger denominator on the right-hand side. Adding site value to the problem shrinks the rate of growth in the value of standing trees as capital assets in comparison to the Wicksell case.

But there is an intuitive explanation, as well. If the landowner solves the optimal rotation problem once,

The Faustmann rule identifies the dynamically efficient forest rotation, maximizing the present value of future net benefits. It takes into account the time value of money, and also site value, the opportunity cost of keeping the land in forest rather than converting it to another use, like farming or residential development.

she has solved it forever, given no change in the basic parameters. That is, $T^\star$ will remain the same in year one hundred as it is in year one. So in addition to delaying the net returns from the current harvest by one year, if she waits one extra year before cutting the trees she also delays the net returns from each future harvest by one year. And each of those delayed harvests has an associated cost—the foregone growth in the value of money in the bank from the harvest over the period of one year. The Faustmann model captures the present value of this change in the timing of all future harvests in S. The perpetually delayed harvests decrease the present value of expected future returns to the land (which before $T^\star$ is outweighed by the marginal increase in the present value of expected future returns that results from greater timber volume). And taking into account this future loss from delaying the harvest shortens the optimal infinite rotation, in comparison to the single rotation. While this characteristic of the efficient rotation is extremely important, it is also quite a subtle point. It is so subtle, in fact, that even though German forester Martin Faustmann described the rule in a paper published in 1849, the substance of his point went unnoticed by forest economists for more than a century.

Efficient Rotation with Nontimber Forest Benefits

Forests offer many benefits aside from the commercial value of their timber. Nontimber forest products and benefits include species habitat, watershed protection, carbon sequestration, and recreational opportunities. How do these nontimber forest values affect the efficient rotation?

When a forest provides multiple nontimber benefits, it can be hard to sort out their effects on the optimal rotation. Imagine, for example, that our landowner's forest stand provides habitat for both the red-cockaded woodpecker (a threatened U.S. species which prefers mature southern forests where trees reach 60 to 180 years of age) and the white-tailed deer (which prefers tender new growth and lots of understory on which to graze). These two uses pull in opposite directions. To understand the general impact of nontimber forest benefits on efficient forest management, we will consider an example with benefits from old-growth standing trees, like the woodpecker habitat described above. Keep in mind, however, that the problem is often multi-dimensional and therefore more complicated.

In figure 7.4, we introduce species habitat value—assumed to accrue only after the stand is sufficiently old—into our forest management problem.[2]

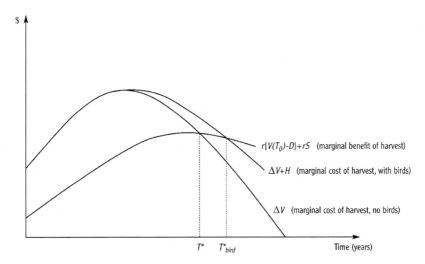

Figure 7.4 The effect of non-timber benefits on the Faustmann rotation. The value of bird habitat provided by old-growth forest represents an additional cost of harvesting timber. As a result, the efficient rotation length increases in this case, from T^* to T^*_{bird}.

We do this by incorporating the foregone benefits of woodpecker habitat as an additional cost of harvesting timber (since the benefits of harvesting are captured in the commercial value of the timber). Thus, the *social* marginal cost of harvesting timber includes both the foregone potential growth in timber volume from waiting one more year (ΔV), and the foregone benefits of bird habitat (H).

The effect is to increase the optimal rotation period. If the landowner is managing her forest efficiently, she will now let the trees stand longer in each rotation than she would have in the absence of bird habitat. In figure 7.4, we represent this as an increase in the optimal rotation from $T^\star$ to $T^\star_{bird}$. In fact, if the value of woodpecker habitat in old-growth forests is large enough, the social marginal cost of harvest curve may never intersect the curve that describes the marginal benefits of harvest. In other words, the optimal rotation may be infinite, meaning that it would be efficient to never harvest certain stands.

The benefit valuation techniques described in chapter 3 can be used to determine the effect of nontimber forest benefits on optimal forest rotation. For example, in the coastal forests of British Columbia and boreal forest of

Nontimber Forest Benefits: The Spotted Owl Controversy

In response to new biological information on the habitat requirements of the northern spotted owl, the U.S. Forest Service was directed in the mid-1980s to issue a revised management plan for the Pacific Northwest Region. The "preferred management alternative" presented in this plan generated a large amount of media coverage and controversy. There were lengthy arguments for and against measures to set aside old growth forests for owl habitat. By the late 1980s, there was open disagreement among federal agencies regarding the management plan, and environmental advocates were attempting to block timber sales from old-growth areas. Court rulings alternately halted and permitted timber sales, and most decisions were appealed by either the logging industry or environmental groups or both. By March 1989, twenty-five timber mills in the region had shut down, causing many workers to lose their jobs. In 1990, the owl was listed as a threatened species under the Endangered Species Act.

These are difficult trade-offs: jobs, threatened species dependent on old-growth forest, recreation, cheap timber for construction and other purposes. A number of economic analyses strove to make these trade-offs transparent in the form of benefit-cost analysis. One group of economists estimated the benefits to U.S. residents of preserving old-growth forests in the Pacific Northwest as habitat for the spotted owl.[3] Their benefit estimate, determined through contingent valuation to be in the neighborhood of $1.5 billion, exceeded the U.S. Forest Service's estimate of owl preservation costs (including foregone timber harvests), which ranged from $500 million to $1.3 billion. Another study also showed total benefits of owl habitat protection greatly exceeding total costs (their most conservative estimated ratio of total benefits to total costs was 3:1, and the estimated ratio most favorable to preservation was 43:1).[4]

northern Alberta, Canada, the Faustmann rotation increases by approximately 20 percent, on average, when carbon sequestration benefits of standing trees are considered.[5] Another analysis of a forested watershed in Victoria, Australia, makes a similar type of calculation.[6] They account for the value of both watershed protection and carbon sequestration, finding that under some reasonable assumptions with respect to the value of water, carbon, and other factors, the particular forest should never be harvested.

Public Goods, Property Rights, and Deforestation

The Food and Agriculture Organization (FAO) of the United Nations estimates that 14.2 million hectares of tropical forest were lost each year between 1990 and 2000, a rate of loss offset only slightly by new plantings and conversion of land to forest plantations in these regions. Given the richness in biodiversity of tropical forests, deforestation, particularly in the global south, has been an ongoing issue of concern to environmental advocates and policy makers. As northern countries have increasingly regulated the extraction of timber, especially from old-growth forests, partially due to greater recognition of nontimber forest values, the resulting restrictions in global timber supply and increases in global timber prices have increased the incentives for deforestation in tropical regions (which are comparatively less heavily regulated).

To the extent that the world's forested lands are in private hands and markets are complete, some of this observed deforestation is actually efficient use of a scarce resource: namely, land. In previous chapters, we argued that private owners of nonrenewable natural resources in competitive markets faced powerful incentives to maximize the present value of net benefits from their resources, like oil wells and coal mines. In the same way, private landowners face powerful incentives to maximize the present value of the net benefits they receive from their land. So, for example, farmers think carefully about the profit-maximizing mix of crops to grow in a given year, within certain constraints. In the long term, they also think carefully about whether farming, itself, is the most profitable use of their land, or whether they should instead consider selling their land to residential developers. Farmers in developing countries, likewise, make similar decisions about converting forested land to agriculture, although the context is different in important ways.

While nonrenewable resources lie, in large part, in private hands, forests exhibit a wide variety of ownership regimes. For example, in 1991, federal and state governments owned approximately 49 percent of U.S. forested lands, and 51 percent were privately owned (10 percent by the forest industry and 41 percent by small farmers and other landowners).[7] In developing countries, more than 80 percent of forested lands are publicly owned.[8] In addition, many nontimber forest benefits are public goods, so we would expect private landowners to rotate their trees at a rate that provides less than the efficient quantity of some of these services, like species habitat, watershed protection, and carbon sequestration. These two factors—the nature of property rights and the prevalence of public goods among forest services—frame the discussion of deforestation from an economic perspective.

Well-defined property rights are particularly important for the efficient management of forest resources, because optimal rotation periods for some species can be very long. Imagine trying to ensure that a valuable hardwood left standing today will be there for the efficient forester in fifty years, in a country where the revenue from capturing the hardwood can feed a hungry family for months at a time. Even where well-intentioned governments intervene in markets to establish property rights to forested lands, the incentive to poach trees in such countries is very high. Efficient forest management hinges critically on the ability to regulate the capture of forest resources. Where this is not possible, we observe rotations that are too short, from the perspective of efficiency.

Economists have provided empirical evidence regarding the influence of property rights on deforestation rates. Studies of the Amazon basin in Brazil have shown that possession of land title leads to longer rotation periods and increased efforts at reforestation and conservation by small landholders.[9] Land title holders in Brazil are also less likely to participate in timber markets altogether—they are less likely to sell trees for a living than to use forested lands for other purposes. In developing countries, there is a strong relationship between deforestation rates and factors (like political instability) that indicate uncertainty over property rights.[10]

Economists have also shown the importance of protection and enforcement of property rights, in addition to their establishment.[11] In Brazil, while the possession of land title reduces the incentive to deforest one's land, the positive effect of land title is eroded where title-holders face a continued risk of fire contagion from neighbors clearing land for agriculture. Social institutions can mitigate this effect. For example, if landowners are aware of the timing of neighbors' fires, they can take preventative measures. Where such coordination occurs, title-holders maintain longer rotations and better conservation and reforestation practices.[12] Thus, property rights alone cannot guarantee the maximization of net benefits from forested lands, but they are an important step in the direction of more efficient management.

Fisheries

Forests exhibit a range of property rights regimes, from private ownership to open access. Unless we include aquaculture in our analysis, the same cannot be said of fisheries. In fact, most fisheries are not owned by any party, and many are open access; thus, completeness of markets will be an integral part of our fish story. As in the case of forests, we will first take a look at a simplified version of fishery biology, and then add the economic dimension to determine the efficient quantity of effort to exert in harvesting fish.

Logistic Growth

To represent the biological side of our bioeconomic model of a fishery, we employ a commonly used growth model—the Schaefer logistic model for growth of a species population—that describes incremental growth in a fish stock as a function of the size of stock (usually measured in tons of biomass).[13] The general form of the logistic function is as follows, where X is the fish stock, r is the fish species' intrinsic growth rate, and K is environmental carrying capacity:

$$F(X) = rX(1-\frac{X}{K})$$

The logistic curve is symmetric and bell shaped (see figure 7.5). To the left of the curve's peak, the annual growth in stock is increasing in the size of the stock (bigger stock leads to faster growth), although this is happening at a decreasing rate. When we reach the peak, the annual rate of growth is maximized (at X_M). On the right side of X_M, the rate of annual growth in stock is decreasing in stock size—more fish mean less growth, as the larger population begins to result in crowding and competition for food, for example. At the far right-hand side of the logistic curve, we reach the carrying capacity of this fishery—K. This is the fish population that would persist in the absence of any outside perturbation; mortality is exactly offset by new births.

Bioeconomic Model

Now we introduce fishing activity. Harvesting fish, like extracting oil, is fundamentally a dynamic problem. The population left in the fishery tomorrow will depend on what is happening today, such as the size of the fish stock, environmental conditions, and human fishing effort. Solving the full dynamic problem turns out to involve some very complicated mathematics. To keep things simple, we'll focus instead on a "steady-state

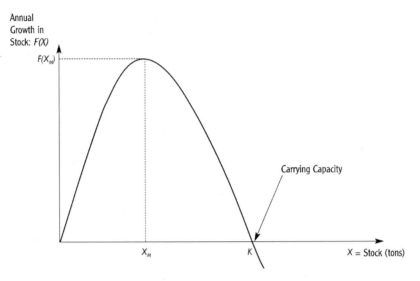

Figure 7.5 The logistic growth curve in the model of a fishery.

model"—that is, on what happens in a fishery in long-run equilibrium. This approach will still give us a great deal of insight into the economics of fishing and the inefficiency of open access.[14]

In the steady state, the fish population remains the same from period to period, thus the fishing harvest is equal to net growth of the stock. Notice that any harvest level would be sustainable if the harvest rate were equal to the growth rate of the fish stock. Such a harvest could be sustained forever, and the underlying population size would remain constant. For any level of the stock (X), $F(X)$ is equal to the rate of annual growth in the stock, and it is also equal to the maximum sustainable long-run level of harvest for that stock. Thus in the steady state, the logistic curve is also a sustainable yield curve.

Looking at figure 7.5, we can expect to maintain any population size along the horizontal axis between zero and K simply by harvesting the number of fish each year equal to the natural change in population from the last year. For example, X_M is known in biological fisheries models as the stock that results in *maximum sustainable yield*. A stock of this size maximizes the average level of growth; hence it also maximizes the sustainable (non-stock-reducing) yield—that is, the largest catch that can be perpetually maintained, $F(X_M)$. If the fishing harvest were to reduce the stock beyond this point, the fishery would be biologically overfished.

In order to understand the relationship between fishing effort and the returns to fishing in the steady state, we need to make an important assumption. We assume that the yield per unit of fishing effort is proportional to the size of the fish stock— more fish mean a greater return per unit of fishing effort.[15] Given this assumption, the relationship between the logistic growth curve we have already drawn and a curve describing the steady-state returns to fishing as a function of the level of effort is quite straightforward—they look very much alike. See the appendix to this chapter for a formal derivation of the yield-effort function if this seems like too large a leap.

In biological fisheries models, the stock that maximizes the average level of growth also maximizes the sustainable (non-stock-reducing) yield—that is, the largest catch that can be perpetually maintained. If the fishing harvest were to reduce the stock beyond this point, the fishery would be biologically overfished.

In figure 7.6, we graph the yield-effort function $Y(E) = 10E - E^2$. We represent fishing effort (any unit will do; here we use the number of boats) on the horizontal axis. With the level of fishing effort increasing from left to right, the fish stock is increasing from right to left, simply because fishing effort reduces the stock. Thus, at the origin, effort is equal to zero and the stock is equal to K, carrying capacity or natural equilibrium, the level of stock that will prevail in the absence of fishing. The vertical axis in figure 7.6 measures the returns to fishing effort. We could continue to measure returns in tons of fish, but we find it more useful to proceed in dollars. To convert fish to dollars, we need only to know the price of fish; here, to keep things simple, we assume it to be one dollar per ton. The hill-shaped curve in figure 7.6 gives us the total revenues we can expect from fishing at varying levels of fishing effort in the long run. Notice that after the curve peaks, effort continues to increase while returns decline. We have also drawn in the total cost curve, assuming that the marginal cost of fishing effort is three dollars per boat.

Efficient Fishery Management versus Open Access

To decide how much effort we should put into this fishery, we compare benefits and costs. To maximize the value of this resource to society, the fish-

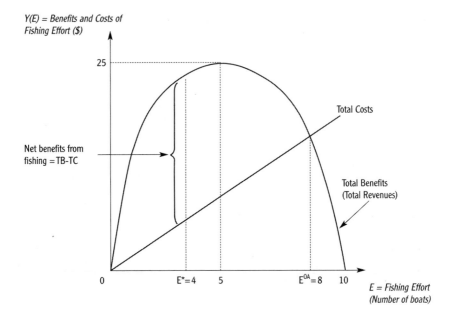

Figure 7.6 Efficiency versus open access. The efficient level of fishing effort, E^*, sets marginal cost equal to marginal benefit. In the open-access equilibrium, total costs and benefits are equal, resulting in a much higher level of effort, E^{OA}.

ing effort should maximize the rents from fishing (the difference between total benefits and total costs). Rents in the fishing model are like scarcity rent in the nonrenewables model—in both cases, rents are what we get back from a scarce resource above and beyond the cost of extraction or harvest. We can locate the rent-maximizing level of effort using figure 7.6—it is the point along the horizontal axis at which the distance between the total benefits curve and total cost curve is maximized (E^* in figure 7.6).[16] In our example, this occurs when four boats are employed in the fishery. Fishing effort beyond this point is *economic* overfishing. Notice that the fishery is economically overfished before it is biologically overfished. This will always be true because fishing is costly. If the marginal cost of fishing effort were zero, the efficient level of effort would be equal to the level of effort that would preserve a stock that ensures maximum sustainable yield.

Is the level of fishing effort that we observe in real-world fisheries economically efficient? By now, you have probably guessed that it is not. We have discussed this problem as if fisheries had walls and gates, and were either privately owned or staffed by a social planner who had the power to limit the number of boats admitted. While there are privately owned fish farms in many countries, deep-ocean fisheries (those beyond individual nations' two-hundred-mile exclusive economic zones) are characterized by open access.

Fishing effort beyond the point at which marginal benefits are equal to marginal costs is economic overfishing.

Under open access, boats will continue to enter a fishery as long as there are profits to be made (as long as rent is positive). At the efficient level of fishing effort identified in figure 7.6, profits still remain for further boats. This will be true as long as the total benefits curve lies above the total cost curve—that is, past the efficient number of boats, past the number of boats that result in the fishery's maximum sustainable yield and total returns begin to decline, to the point at which total benefits are exactly equal to total costs. In our model, in the absence of private ownership or regulation, fishers will continue to enter until there are eight boats on the fishery, twice as many as the efficient number (E^{OA} in figure 7.6).

At the open-access equilibrium level of fishing, the rents to fishing are completely dissipated. At first glance, the absence of such rents might seem to be the result of healthy competition in a well-functioning market. But that overlooks a crucial aspect of this problem. When eight boats enter this model fishery, they generate sufficient returns to cover the costs of their harvest, including things like fuel, gear, and depreciation of fishing vessels. But there is nothing left to cover the depletion of an important resource—the fish!

Here again, actors in a market generate costs (fish depletion) that are external to the transactions between buyers and sellers. No individual fisher has an incentive to account for the externality of depletion of the fish stock. Like our conscientious farmer who leaves a gallon of water in a nonrenewable groundwater aquifer, hoping to save it for tomorrow, the conscientious fisher who leaves a succulent tuna in the sea for tomorrow will quickly be put out of business by her competitors, who indulge in no such collectively minded behavior. The result is an inefficient race to fish, with too many boats pursuing too few fish. This depletion of the resource resulting from a conflict between collectively beneficial and self-interested behavior is a classic example of the tragedy of the commons that we discussed in chapter 5.

Common Property Is Not Open Access

Having read the preceding discussion, it may seem as though we are imply-ing that all shared resources, from fisheries to groundwater aquifers, are doomed to inefficient overexploitation. This is not the case, however. Al-though the terms "open access" and "common property" are often used in-terchangeably to describe shared resources, they are really very different property rights regimes. Open access resources, by definition, lack any re-striction on who can use the resource or how much they can extract. In contrast, common property resources, though shared among a group (for example, a village), may be governed by formal or informal institutions—ranging from explicit rules to informal social norms.[17] Importantly, many common-property arrangements steer clear of the extreme overexploita-tion typical of open-access resources.

To get a feel for the difference between open access and common prop-erty, let us consider two very different social situations. In the first, you go out to dinner at a local restaurant, where rather than charging each diner the precise cost of her meal, the manager has decided to split the evening's total billings evenly among all of the guests in the dining room. You will pay a bill that amounts to the average of the entire restaurant's tab. How will your ordering behavior differ from the standard situation in which you pay for your own food and drink? Most of us would be more likely to order that second glass of wine, or coffee and dessert, when splitting the bill with a crowd of strangers.

In the second scenario, you go out to dinner at the same restaurant with a small group of close friends. You agree to split the bill evenly. Do you order that second glass of wine, as you did when you were splitting the tab with a large crowd of strangers? You might be somewhat less constrained in your ordering than you would be if you were going to pay the full cost of your order yourself. But you are likely to be somewhat more constrained than you were in the first scenario. After all, your friends might think very poorly of you if you stuffed yourself at their expense.

Open access resources lack any restriction on who can use the resource or how much they can extract. In contrast, common property resources, though shared among a group (for example, a village), may be governed by formal or informal institutions.

The first scenario resembles open access, the second common property. Successful common-property arrangements rely on the self-regulating (and self-enforcing) capacity of resource users—various incentives not to act opportunistically. Individual interest can be constrained by formal or informal institutions, like the social norms observed by a group of friends in a restaurant.

There is an intuitive link here to the Coase Theorem, which we will address in chapter 8. By limiting harvests, fishers can increase the total returns from a shared fishery even as they exert less effort. The simple fact that potential rents exist and will be dissipated in the absence of some kind of disciplined use may provide a powerful incentive for imposing such discipline. Of course, even successful common property arrangements, while they may avoid the degree of depletion observed under open access, may not result in efficient use. Thus, a common-property resource may be economically but not biologically overexploited.

Caveats Regarding the Steady-State Approach

Before we use the intuition from this simple model to analyze what we see happening in real-world fisheries, we pause to discuss three important details that are lost in the translation from a full dynamic bioeconomic fishing model to the steady-state model we have just described.[18] First, the steady-state analysis does not account for the time value of money. In a fully dynamic model, we would discount future costs and benefits of fishing effort, thereby arriving at an efficient level of fishing effort that is somewhat higher than that we identified using the steady state, but still lower than the open-access equilibrium. Second, in the dynamic analysis, it becomes clear that even the open-access equilibrium may not be sustained in some fisheries. In these cases, fisheries may be so intensely exploited that the stock may collapse entirely. Third, without going through the dynamic model, it is impossible to demonstrate the strong parallels between the choice of a dynamically efficient harvesting policy and the choices of the optimal extraction rate of a nonrenewable resource, or the optimal rotation length for a forest. Most of the intuition we developed in those cases carries over to the fisheries problem. In particular, efficient fishery management has a strong Hotelling-like feel, in which the rate of return of the resource (fish in the sea) must equal the interest rate.

Open Access versus Common Property: Beaver Hunting in James Bay

An example of the difference between open access and common property regimes in the realm of natural resource management is that of James Bay, Quebec, where hunters traditionally have used resources communally.[19] The local Native American peoples have a rich heritage of customary laws regulating beaver hunting—beaver in the region are an important food species and, since the start of the fur trade in the region in the late 1600s, a commercial species, as well. Beavers are vulnerable to depletion because colonies are easily spotted. Historically, a common property arrangement with senior hunters and their families serving as "stewards" of specific territories ensured sustainable use.

In the 1920s, a large influx of outsiders arrived in response to high fur prices. Native communities lost control over traditional territories, and a race to hunt ensued. All trappers (native and nonnative) contributed to the resulting tragedy. Beaver populations reached an all-time low in 1930, and conservation laws were enacted, banning outsiders from trapping. The traditional native family territories were recognized, and customary laws became enforceable. The return to a common-property arrangement generated productive harvests again after about 1950.

Economics and Real-World Fisheries

According to the FAO, of the major marine stocks or species groups for which information is available, about 18 percent are biologically overexploited, meaning that they are fished beyond the point at which the stock is able to maintain maximum sustainable yield; another 10 percent have been significantly depleted and are in danger of collapse.[20] The collapse of the cod, flounder, and haddock fisheries on the Grand Banks off the coasts of New England and Canada are one commonly cited example. In the late 1400s cod were so abundant that they could be pulled from these waters in weighted baskets.[21] Groundfish landings peaked in the Northwestern Atlantic in 1965 at 2.6 million metric tons; despite significant improvements in fishing technology and substantially increased effort, landings in 1994 were less than 350,000 metric tons.[22]

Can we draw any direct links between the inefficiency of open access in theory and the numerous examples of fishery collapse in the real world?

Open access: the Gulf of Mexico shrimp fishery

The Gulf of Mexico shrimp fishery is one of the most valuable fishery resources in the United States. It is an open-access resource and has suffered from over-fishing.[25] Landings increased from an annual average of 193 million pounds between 1950 and 1980 to 243 million pounds between 1980 and 1990. The number of vessels more than doubled between 1965 and 1988, while average landings and revenue per vessel declined. In addition to biological overfishing and the dissipation of rents, overcapitalization in the Gulf shrimp fishery has resulted in significant by-catch and discard of other species, as well as harm to endangered marine turtles. One study of the Gulf of Mexico estimated that in 1988, the annual shrimp harvest could be maintained at the same level by one-third of the existing shrimp-trawling fleet.

A number of studies have attempted to estimate the efficient level of fishing effort in various fisheries, and to compare actual rates of effort with efficient rates. A regional study of a fishery in the Bering Sea and Aleutian Islands concluded that the optimal level of effort would allow 24 factory trawlers and 44 to 50 catcher vessels in the fishery; 140 vessels were operating in the fishery at the time.[23] U.S. fisheries that have experienced near-total depletion as a result of open access and resulting overcapitalization include those for Atlantic halibut, Northern California sardine, Atlantic Ocean perch, and West Florida sponge.[24]

Other Important Considerations: Subsidies and Externalities

We have two remaining issues to explore in our discussion of fishery economics: subsidies and by-catch. Fishing industry subsidies can worsen the open-access problem and accelerate fishery collapse. Subsidies are a widespread attempt of governments to support those employed in fisheries that have become economically marginal through open-access or other forces. They can take many forms, including:

- Direct income support
- Price supports
- Reductions in the marginal cost of fishing effort, like fuel tax exemptions

- Subsidies of capital equipment, including low-interest loans and loan guarantees
- Inefficiently low (or even zero) charges for fishing in public waters
- Subsidies to shipbuilding, ports, and fish processing facilities

Where subsidies are in place, even in the absence of open access, the equilibrium level of effort in a fishery will be higher (and stocks will be smaller) than the efficient level. In the early 1990s, the FAO estimated that the world's 3 million fishing vessels had $92.2 billion in annual operating costs (excluding vessel depreciation, debt service, and return on investment), but brought in only $70 billion in gross revenues.[26] The shortfall was made up by government subsidies. A more careful study by the U.S. National Marine Fisheries Service estimates that world governments spend $15 to $20 billion per year on fishing subsidies.[27]

To understand the impact of subsidies on the equilibrium level of fishing effort, let us take, for example, a specific type of subsidy—a fuel tax exemption, which drives down the marginal cost of fishing effort. In figure 7.6, a decrease in the marginal cost of fishing effort would pivot the total cost curve downward from its stationary point at the origin toward the horizontal axis—its slope would be less steep. As a result, the open-access equilibrium level of fishing effort would be pushed even further to the right. Because we have chosen to measure effort in terms of the number of boats, in our example additional boats will enter the fishery in the presence of a subsidy, resulting in further depletion of the stock. Keep in mind, however, that subsidies reduce fish stocks in two ways; they lower the cost of fishing for boats currently operating on a fishery (potentially increasing current boats' number of fishing days), and they make fishing profitable for previously marginal boats (the effect we describe above).

The second remaining issue in our discussion of fish has to do with externalities. We have already discussed an important externality—the "depletion externality" that is present in open-access fisheries. Another externality associated with fishing is that of by-catch, the unintended capture of non-target fish and marine mammals. By-catch imposes a cost on society—the loss of these innocent bystanders—that in the absence of regulation is external to the decisions of individual fishing boats. Thus, where by-catch occurs, the equilibrium level of fishing effort (even in the absence of open-access) will be higher than the efficient level. One example is the dolphin-tuna controversy that erupted in the late 1980s and resulted in the labeling of "dolphin-safe" tuna. Between 1960 and 1972, an average of one hundred thousand dolphins were killed each year in by-catch incidents by the U.S. tuna fleet alone.[28] Regulation in the form of the Marine Mammal Protection Act required fishers to take measures to decrease dolphin mortality beginning in 1975; current mortality levels are approximately five thousand per year.[29]

Conclusion

The models we have discussed in this chapter and the previous one are members of a class of problems regarding the economics of natural resource management. By nature, as we have seen, the management of a natural resource—whether nonrenewable (like oil) or renewable (like forests or fish), is inherently a dynamic problem. Economic analysis provides a framework for decisions regarding the allocation of scarce natural resources over time—just as it provides a framework to analyze the allocation of other scarce resources among firms and consumers. We have discussed natural resource stocks as capital assets: Goods that provide returns over time, either in their natural state (growing trees, oil in the ground), or upon extraction or harvest.

The bottom line is that efficient management requires that extractors and consumers of natural resources face the "right" prices to enjoy these activities. In some cases, markets acting alone will get prices right. Private owners of oil reserves in competitive markets, for example, can be expected to take scarcity into account in deciding how to allocate their extraction over time. In other cases, markets require some type of intervention to ensure efficient prices. If forests generate public goods, such as species habitat, private owners of forested lands may underestimate the social cost of the timber harvest, and thus harvest trees too quickly. If the cost of harvesting a fish from a deep ocean fishery includes the labor and capital used to capture it, but not the value of the diminished fish stock, fisheries will be depleted too rapidly.

Later in the book, we will come back to the question of "getting prices right" in markets for natural resources—ensuring that all of the externalities associated with resource extraction and consumption are reflected in market prices. In chapter 8, we will discuss government policies that can mitigate the open access problem and other externalities. And in chapter 10, we will see how one such policy—known as "tradable individual fishing quotas"—has been implemented in the real world.

Appendix

The Yield-Effort Function for a Fishery

We begin with a "production function" for the fishery—a function that describes the relationship between the steady-state yield, the stock, and the level of human fishing effort:

$$Y_t = H(X_t, E_t)$$

where:

Y_t = yield at period t
H = harvest
X_t = stock at period t
E_t = fishing effort at period t

The change in the fishery's biomass from period to period will be equal to the natural growth in stock, less the yield from fishing (recall from the text that the growth in stock is equal to $F(X_t)$):

$$X_{t+1} - X_t = F(X_t) - Y_t$$

Now we can substitute our production function into this equation, describing the change in stock from period to period as a function of the initial stock, and the harvest (which is, itself, a function of the initial stock and human fishing effort):

$$X_{t+1} - X_t = F(X_t) - H(X_t, E_t)$$

In order to proceed with our simplified analysis, we rid ourselves of the time dimension by starting from the steady state, at which the stock is unchanged from period to period, so that $X_{t+1} = X_t = X$. Thus, it must be true that $F(X) = H(X,E)$; we have just restated our assumption from the text that in the steady state, the level of growth in the stock will be exactly equal to the harvest from fishing.

The key economic variable of interest here is the level of fishing effort to exert in the fishery—it is the only variable in the system that is entirely within human control. Thus, we would like to establish the relationship between the fish stock and the level of fishing effort: $X = G(E)$. If we

The steady state yield-effort function describes the long-run relationship between fishing effort and the returns to fishing.

substitute this into our production function (above), we obtain the relationship between yield and fishing effort: $Y=H(G(E),E)=Y(E)$. This is the steady state or sustained *yield-effort function*, which describes the long-run relationship between fishing effort and the returns to fishing. If we apply effort level E each year on an ongoing basis, the sustained annual yield we will obtain is $Y(E)$.

Let us now substitute actual functions for $F(X)$ and $Y(E)$. To describe the growth in stock, we will use the logistic function from the text:

$$F(X) = rX(1 - \frac{X}{K})$$

For the yield-effort relationship, we will use the very common constant-returns-per-unit-effort (CPUE) function: $Y = qXE$. In the steady state, recall that yield will be equal to the growth in stock, $F(X) = Y$. If we use this fact to solve for steady-state yield as a function of effort, we obtain:

$$qXE = rX(1-\frac{X}{K})$$

$$X = K(1-\frac{qE}{r})$$

$$Y = qE(K(1-\frac{qE}{r}))$$

$$Y = qKE(1-\frac{qE}{r})$$

Note the similarity in functional form between the yield-effort function and the logistic growth function. In this steady-state model, the graphs describing $F(X)$ and $Y(E)$ look very much alike. To understand why this is true, let us derive the relationship between the fish stock (X) and fishing effort (E), solving for E in terms of X.

If $Y(E)=qXE$, and in the steady state, $Y(E)=F(X)$, then we can also say that:

$$E = \frac{Y(E)}{qX} = \frac{F(X)}{qX}$$

$$E = \frac{rX(1-\frac{X}{K})}{qX}$$

$$E = \frac{r}{q}(1-\frac{X}{K})$$

So effort is the difference of some constant (r/q) and that constant times the stock divided by carrying capacity (X/K). This explains why the graphs look quite the same in the steady-state model. Both are hill-shaped parabolas, and they have the same maximum point ($K/2$ for $F(X)$ and $r/2q$ for $Y(E)$).

8

Principles of
Market-Based
Environmental Policy

Is government intervention in the environmental arena needed at all? After reading the earlier chapters, you may think that the answer is self-evident. After all, in the absence of government policies we have seen that private firms and individuals may impose negative externalities on other members of society, and will fail to provide efficient amounts of public goods.

Nonetheless, a strong argument can be made that individuals or firms—at least in some cases—can solve externalities on their own, through private bargaining. This argument, due to Ronald Coase, is the starting point for this chapter; it offers a presumption against government intervention that we must consider carefully. After we have satisfied ourselves that government still has a vital role to play in addressing environmental problems, we go on to discuss the various types of regulatory policies (or "policy instruments") that governments can employ. These include prescriptive regulations that mandate certain actions at the level of individual firms, as well as more flexible market-driven approaches. As we shall see, these market-based policies—which include taxing emissions and creating markets in pollution—are appealing from an economic perspective, because they directly address the market failure at the root of environmental problems. Although a main theme running through our discussion is the common logic underlying emissions taxes and cap-and-trade programs, we will close the chapter by considering when it makes sense to use one rather than another on the grounds of economic efficiency.

The Coase Theorem

We opened this chapter by asking whether government intervention was necessary to correct negative externalities such as air pollution. This topic is the basis of a famous debate in economics—one whose central protagonists faced off not in person, but in print. On one side was Arthur Pigou, author of a classic treatise on public goods, *Economics of Welfare*, first published in 1920. Pigou used economic theory to argue for government intervention in the economy—a sharp departure from Adam Smith's Invisible Hand, but one that soon became conventional wisdom among economists. Consider (wrote Pigou) a railroad running through woodland. Sparks from the railroad threaten to start a fire along the tracks, destroying the woods. The railroad, however, will ignore this adverse effect, and proceed to run as many trains as will maximize its profit. Pigou argued that this sort of external effect called for government action. To solve it, he suggested making the railroad liable for damages—either by requiring the railroad to compensate the landowner, or simply by levying a tax.

Four decades later, Ronald Coase challenged this view—first in passing in an analysis of radio and television broadcasting, and then head-on in a classic article titled "The Problem of Social Cost," which helped him win the Nobel Prize in Economics many years later.[1] Coase argued that under certain conditions private bargaining between the railroad and the landowner would result in the same outcome—the same number and speed of trains, say—regardless of whether the railroad is liable for damages or not.

How might private bargaining overcome negative externalities? Coase used the example of a rancher raising cattle next to a farmer growing crops. Without fences to protect the crops or to contain the cattle, the livestock would stray and eat the crops. The Pigouvian remedy would be to tax the rancher for the damage. In that case, the rancher would pare back his herd or build a fence in whatever way maximized the gains from cattle raising, net of the damages caused. This would evidently be the efficient outcome. Coase's key insight was that the identical outcome would arise without government intervention. If the rancher is permitted to let his cattle roam with impunity, then the farmer will build a fence (or pay for a reduction in the herd) if and only if the cost of doing so is less than the avoided damage to the crop. Left to their own devices, therefore, the farmer and the rancher would still reach the efficient outcome.

In the jargon of economics, the Coase Theorem states that the allocation of property rights (i.e., whether the rancher has a property right to let his cows roam as they please, or the farmer has a right to a cow-free cropfield) has no bearing on how economic resources are used (i.e., whether the fence is built). Of course, how those property rights are allocated matters to the farmer or the rancher, because it affects the distribution of income and wealth; but it does not matter from the point of view of society as a whole.

So striking was Coase's argument that when he originally proposed it, even the economists at the University of Chicago—the citadel of free-market economics—thought that he must have erred. They invited him to Chicago to make his case at a dinner party hosted by Aaron Director, chair of the formidable Chicago economics faculty. As recounted later by George Stigler (one of three eventual Nobel laureates in attendance), "In the course of two hours of argument the vote went from twenty against and one for Coase to twenty-one for Coase." By dint of persuasion, Coase had convinced his colleagues that the case for government intervention to solve negative externalities was weaker than Pigou had argued. If government actions do not promote efficiency, then why should government get involved at all? After all, government policy is costly, and may have unintended consequences that make matters worse rather than better.

Hidden in our discussion so far, however, has been a crucial assumption: that bargaining is easy and inexpensive, and that deals are easy to enforce. In the case of neighboring landowners, that is plausible enough. But what of the case in which soot from a factory settles over an entire town? Finding all affected individuals will be difficult. Determining the true damages will be nearly impossible (since if the factory is to pay compensation, individuals will have strong incentives to overstate their damages). And in the case where the polluter is not liable (so that efficiency might require the individuals to pool their funds and pay the factory to install pollution control equipment), the individuals themselves will face a classic collective action problem: Each person, realizing that their contributions benefit everyone else, will seek to free ride on the efforts of others. In such a case, the costs to the town and factory of reaching and enforcing a bargain—

The Coase Theorem states that private bargaining will overcome negative externalities, without the need for government intervention, regardless of how property rights are allocated.

what economists call *transactions costs*—are likely to be insurmountable. In this case, the liability rule will matter for efficiency after all. A factory that faces a tax on pollution will curtail its operations or install equipment to limit emissions; but the townspeople on their own are unlikely to be able to pay the factory to do those things. As a result, the assignment of property rights will affect how much pollution control is done, not just who pays for it. For the Coase Theorem to hold, therefore, transactions costs must be negligible.

But of course transactions costs are not negligible in the real world.[2] As in the example above, transactions costs will generally be large when a large number of people suffer the damages from an externality. They will also arise when large numbers of firms or individuals contribute to the problem, or when causation is difficult to establish, or when information about damages is not widespread, or when firms or individuals act strategically in bargaining situations. In other words, transactions costs are ubiquitous in the environmental realm.

Why, then, should we bother with the Coase Theorem at all? One reason can be explained by way of analogy with physics. Much of classical mechanics takes place in a vacuum. "But everyone knows that we don't live in a vacuum," you might imagine someone saying. "How can this be relevant to the real world?" The answer is that knowing what would happen in a vacuum is a useful starting point for understanding what happens in the real world when you start to incorporate friction, air resistance, and so on. In the same way, the Coase Theorem can be thought of as a kind of "economics in a vacuum" result. By describing what would happen if private bargaining worked smoothly, it allows us to appreciate the importance of taking real-world frictions into account.

A second reason for considering the Coase Theorem concerns what it can teach us about the design of environmental policies. The ubiquitous presence of transaction costs provides a strong justification for government regulation, running counter to Coase's faith in private bargaining. Nonetheless, as we shall see, Coase's key insight—that the clear assignment of property rights may help resolve negative externalities—helped inspire the "cap-and-trade" policies that have been very successful in reducing pollution.

Finally, in some cases of real-world interest, the number of parties is small enough, and the private incentives are large enough, that Coasean bargains can be struck, after all. The accompanying text box discusses just such a case that took place in the French countryside.

Perrier and Vittel: Paying Farmers to Change Their Agricultural Practices

Perrier and Vittel are two leading mineral water companies based in France, now part of the Nestlé Waters Group (the largest bottled water company in the world by revenue). In the late 1980s, Vittel intiated a program to reduce water pollution in the source area feeding the springs that are the source of its bottled water.[3] As part of this program, the company signed long-term contracts of up to thirty years with dairy farmers in the watershed. The farmers agreed to adopt less intensive faming methods in order to reduce agricultural runoff of herbicides and other pollutants. In return, Vittel paid each farmer roughly $230 per hectare per year for a seven-year period—adding up to about $155,000 for the average farm. The company also provided free technical assistance and paid for new farm equipment and construction. When Vittel purchased Perrier in 1992, it applied a similar model to the Perrier springs in southern France, where the program introduced organic farming methods on over five hundred hectares of vineyards and wheatfields.

Although sometimes cited as an example of an "ecosystem services market," this case is more aptly described as a straightforward illustration of the Coase Theorem. Since Perrier and Vittel bottle the water and sell it, they capture sizeable benefits from the improvement in water quality. It seems reasonable to assume that the costs of the program were much less than the costs of alternative methods of cleaning or filtering the water.

At the same time, there may well have been positive spillovers from the private transaction. To the extent that the change in agricultural practices improved water quality throughout the watershed (and not just in the water coming out of the privately owned springs), Perrier and Vittel effectively provided a public good as a by-product of their Coasean bargain.

The Array of Policy Instruments

Despite Coase's concerns, many economists do see a role for government in helping to solve environmental problems. The question is, How should it do so?

Let's start by considering the primary ways government might get involved. We use pollution regulation as a motivating example in this chapter. In chapters 9 and 10, we will broaden our perspective to include a range of real-world examples of environmental policies in areas ranging from fisheries to water pricing to solid waste.

Prescriptive Regulation: Technology and Performance Standards

A first set of policies, known as "prescriptive regulation" or "command-and-control," focuses on regulating the behavior or performance of individual factories and power plants. This has been the conventional approach for most environmental regulation, at least until the late 1990s. A *technology standard* requires firms to use a particular pollution abatement technology. For example, the 1977 Clean Air Act Amendments in the United States required new electric power plants to install large scrubbers to remove sulfur dioxide from their flue gases. Alternatively, the government regulator might impose a ceiling on the air emissions (or water effluent) an individual firm can release. This approach, which allows polluters considerable leeway in determining how to meet those emissions ceilings, is known as a *performance standard* (or emissions standard, in the case of air pollution). It may impose a ceiling on total emissions in a period (for example, tons per year), or a maximum allowable *emissions rate* (for example, pounds of pollution per unit of fuel consumed or output produced). While performance standards could vary among firms in theory, in practice regulators have typically established uniform standards. Finally, real-world regulations are often a hybrid of these two approaches, sometimes called "technology-based performance standards." For example, the Clean Water Act in the United States requires individual sources of water pollution to meet effluent limitations that are based on the "best practicable" (or, variously, the "best available" or "best conventional") technology.

Market-Based Policies: Emissions Taxes and Allowance Trading

In contrast, another set of policies—called "market-based" (or sometimes "incentive-based") instruments—incorporate market principles into government policies. Rather than focusing on the technology or performance of individual firms, these approaches are much more decentralized, focusing on aggregate or market-level outcomes such as total pollution.

Market-based instruments can be divided broadly into two categories, depending on whether they work by influencing prices or by limiting quantities. In the previous section, we mentioned a prime example of the first approach: an emissions tax, set by the government and paid by polluting firms on each unit of emissions. (Other writers refer to the same thing as a "fee" or "charge.") Such a tax puts a price on pollution, forcing firms to recognize the social damages from production along with their private costs. In the language of economics, the tax *internalizes* the costs of pollution, and therefore eliminates the externality (or at least the economic inefficiency associated with it). Subsidies represent another price-based approach. Indeed, a pollution tax and an abatement subsidy can be viewed as flip sides of the same coin. Charging a firm ten dollars for each ton of emissions creates the same incentive to cut pollution as paying the firm ten dollars for each ton of abatement.[4]

The second main approach is known as "allowance trading" or "cap and trade." The government first establishes a total allowable quantity of pollution (the "cap") for a group of firms—for example, all the firms in a particular industry. It then allocates allowances (also called tradable permits) to the regulated firms, with each allowance corresponding to one unit of pollution. (For example, in the real-world case of the sulfur dioxide tradeable allowance program in the United States, each allowance corresponds to one ton of SO_2.) At the end of each year, firms must retire one allowance for each unit of pollution they have emitted during the course of that year. Under a cap-and-trade system, firms that find it relatively expensive to reduce pollution will buy allowances from firms that can reduce their pollution at lower cost. In this way, the total amount of pollution is fixed by regulation, but the *allocation* of that pollution among firms—and therefore the amount of abatement that any single firm must do—is left up to the market.

A number of variations are possible on this basic theme. First, cap-and-trade programs may differ in how the pollution allowances are distributed. In principle, allowances can be sold at auction, raising revenue for the government. In practice, the opposition of powerful industry interest groups has made auctions politically unpalatable. Instead, allowances have been handed out for free to existing firms, usually on the basis of past output. Firms that emit more than their allocation must then buy allowances from firms that emit less. Cap-and-trade systems can also allow firms to save their excess allowances in a bank, to be used in a later year, rather than selling them to other firms. Such banking provisions are typically included when the total allowable amount of pollution is ratcheted down over time. Banking is appealing for two reasons: It gives firms greater flexibility in how to comply with the regulation, and it encourages firms to abate even more than required in the early years of the program.

Information-Based Approaches

Emissions taxes and cap-and-trade policies represent the core of what economists call "market-based instruments." On the periphery are two other approaches to promoting environmental protection that share a similar ethos. The first such approach involves government regulations, often known as "right-to-know laws," that require information provision by private firms. A well-known example is the Toxics Release Inventory. Since 1988, manufacturing facilities in the United States have been required by law to report their annual releases of each of over three hundred toxic chemicals. The Environmental Protection Agency (EPA) makes these data publicly available. (If you are curious, you can find out how much pollution is released in your neighborhood by going to www.scorecard.org.)

The second approach includes ecolabeling and certification programs, which provide consumers with information about how a product was manufactured. For example, you can buy produce from organic farms that do not use pesticides; coffee beans from farms that provide diverse bird habitat; paper made with high postconsumer recycled content or without fiber from old-growth trees; or "eco-friendly" household cleansers that substitute natural ingredients for toxics such as chlorine. Ecolabeling and certification programs aim to advertise and verify such eco-friendly claims. Unlike right-to-know laws, they are voluntary rather than mandatory; indeed, many ecolabeling programs are operated entirely by nongovernmental organizations such as the Forest Stewardship Council and the Marine Stewardship Council, while certification is typically carried out by private firms such as Scientific Certification Systems.

Like emissions taxes and allowance trading, but in contrast to conventional command-and-control regulation, these two approaches are decentralized. That is, they aim to influence the behavior of individual consumers and firms by changing the incentives they face, and then letting them make their own decisions about how to respond—rather than by requiring or proscribing certain activities. For this reason, they are sometimes lumped in with market-based instruments such as emissions taxes and cap-and-trade policies.

However, there is an important difference. Rather than setting up new markets (allowance trading) or introducing price signals (emissions taxes), these approaches work essentially by providing information. As a result, the incentives faced by firms are created indirectly (mediated through the behavior of consumers and citizen groups), rather than being the direct result of government policy (as in the case of emissions taxes and allowance trading). This distinction is critical, because as a result these information-based approaches address a different gap in the marketplace. Right-to-know laws and ecolabeling programs overcome the market failure from *asymmetric information*. That is, they tell citizens more about the activities of factories near them, and inform consumers about the characteristics of the products they buy. But we are still left with the underlying problem of externalities and public goods provision.

In sum, while information provision and ecolabeling can help promote environmental protection, their role is fundamentally different—and more limited in scope—than what can be accomplished through mandatory government policies (such as emissions taxes or tradeable fishing quotas) that directly shape the behavior of the firms and individuals with the greatest impact on the environment. For the rest of this chapter, and indeed the rest of the book, we will leave aside these decentralized policies aimed at providing information, and instead focus on market-based policies.[5]

How Market-Based Policies Can Overcome Market Failure

Economics tends to favor market mechanisms to restore market efficiency. To see why this is, it is useful to recall the discussion of market failure in chapter 5. There, we presented three ways of framing market failure in the environmental arena: negative externalities, public goods, and the tragedy of the commons. Each of these ways of thinking about environmental problems points toward an approach to solving them. One natural solution is to get the prices right, by using government policies to make firms and individuals pay for the environmental damage they cause. Once the negative externalities are internalized in this way, they will be incorporated into the prices of goods and services, and market outcomes will again be efficient. We can also think of government policies as filling in for the missing demand for environmental quality—surmounting the sorts of free-riding problems we talked about in the context of public goods. Or we can think of policies as establishing property rights over resources that had previously been open to all—thereby overcoming the tragedy of the commons.

Getting the Prices Right

In chapter 5, we saw that in the absence of government intervention, firms will ignore the external costs of their actions (e.g., pollution resulting from producing steel) when making output decisions. The market outcome is not efficient because the true social marginal cost of production (which includes the marginal damage from pollution) is greater than the private marginal cost.

As a starting point of our discussion of market-based policies, we'll stick with the simple scenario we used in chapter 5. In particular, let's assume that the amount of pollution is directly proportional to the amount of steel produced. Thus the only way to reduce pollution is to make less steel. While this is not realistic, it makes the intuition behind the tax easy to understand. (We'll consider a more general model of pollution control later.)

Figure 8.1 (which is almost identical to figure 5.1) illustrates this example. The dotted (lowest) line represents the marginal damages of pollution; the gray (middle) line is the supply curve, corresponding to private marginal costs; and the black (upper) line represents social marginal costs. As you know already, the efficient point (Q^*) is where marginal benefit equals social marginal cost. The unregulated market outcome (Q^M) yields too much output, and thus too much pollution—with the size of the ineffiency measured by the shaded deadweight loss triangle.

So far, so good. We now ask: Can a tax produce the efficient outcome? (Note that in this simple framework, a tax on steel and a tax on pollution are identical, since steel and pollution are assumed to be produced in the same proportions.) To answer this question, we need to think about how a tax works. In market equilibrium without a tax, supply equals demand—implying that the price paid by consumers (along their demand curve) is the same price received by suppliers (along their supply curve). A tax, however, introduces a gap between supply and demand. The price that the consumer pays is now higher than the price the supplier receives, by exactly the amount of the tax. To take a simple example, suppose that the government levies a fifty-cent tax on every pack of cigarettes. If consumers pay $4.50 for a pack (including the tax), the retailer retains only $4; the government collects the difference.

On a graph, a tax can be represented by a vertical gap between the supply and demand curves. Market equilibrium, given a tax, is now the point at which the quantity supplied *given the net price to the supplier* equals the

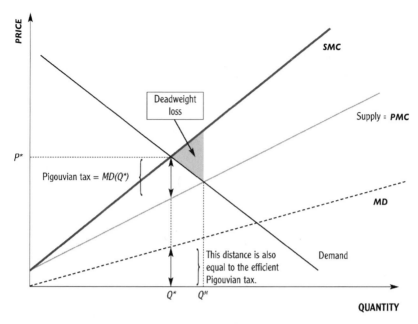

Figure 8.1   The efficient (Pigouvian) tax in the supply and demand framework. The tax equals the marginal damage from pollution at the efficient quantity, Q^*.

quantity demanded *given the price paid by the consumer*. Notice that as long as the supply curve slopes upward and the demand curve slopes downward, both sides of the market (producers and consumers) will share the burden of the tax. That is, the price to consumers will not rise by the full amount of the tax: instead, the price to consumers will rise by somewhat less, while the price to suppliers falls.

Now we can return to thinking about the case of a tax to correct a negative externality. Our goal is to set the tax in such a way that the resulting market equilibrium coincides with the efficient point, Q^*. How to do that? Well, we need to ensure that the tax is exactly equal to the gap (or vertical distance) between the demand and supply curves at Q^*. If so, the wedge created by the tax will be precisely large enough to drive the market to the efficient point.

But notice something important: The gap between the demand and supply curves at Q^* is exactly equal to the marginal damages from pollution at Q^*. This must be the case, since the demand curve intersects the social marginal cost curve at $Q^\star$ (that is the definition of the efficient point), and the difference between the supply curve and the social marginal cost curve is always equal to the marginal damages from pollution. In other words, the optimal tax is precisely equal to marginal damages at the efficient outcome.

Moreover, note that this is the only tax that will achieve the efficient outcome. In particular, a tax equal to the difference between the unregulated market price (P^M) and the price paid by buyers in the efficient outcome ($P^\star$) will not be large enough: that would result in a level of output somewhere between Q^M and $Q^\star$.

This discussion leads to the following observation:

> A tax on pollution equal to the marginal damage at the socially efficient level of pollution will achieve the socially efficient outcome.

The efficient tax is known as a Pigouvian tax, after the economist Pigou (whom we met in the first section). What such a tax does, in essence, is "internalize the externality": It forces the producers and consumers of polluting goods to incorporate the full costs of their actions (including the external costs from pollution and so on) into their output and consumption decisions. With the tax in place, the market outcome will be efficient: No other government intervention (such as telling the firms how much to produce) is necessary.

By correcting the market failure, the Pigouvian tax eliminates deadweight loss. You may have heard or read economists refer to taxes as "distortionary." That is certainly true for most taxes (like sales taxes or income taxes) that raise revenue but interfere with the smooth operation of the market by driving a wedge between supply and demand. In the case of the Pigouvian tax, however, the distortion arises from the *absence* of government intervention. The tax is *corrective*, not distortionary: It eliminates deadweight loss, rather than introducing it. By incorporating the external cost into the price of the good, a tax gets the prices right and restores efficiency.

The efficient (or Pigouvian) tax "internalizes the externality," forcing producers and consumers to incorporate the full costs of their actions into their decisions.

What Is the Optimal Gasoline Tax?

Although we have been discussing pollution taxes on firms (e.g., steel mills), we know from our earlier discussion (chapter 5) that negative externalities can also result from individual actions—for example, driving a car.[6] You are probably familiar with the environmental consequences of automobiles: They are a leading source of emissions of particulate matter and nitrous oxides (which contribute to local air pollution) and carbon dioxide emissions (which help cause global warming). But driving also involves a host of other negative externalities, including traffic congestion and the costs of accidents (that is, the external costs that are not borne directly by the driver).

In a recent paper, the economists Ian Parry and Kenneth Small set out to quantify these various externalities and compute the optimal gasoline tax. They found that the Pigouvian tax reflecting marginal external costs would be eighty-three cents per gallon in the United States. They break that total down among the four externalities as follows: six cents for carbon dioxide emissions, eighteen cents for local air pollution, thirty-two cents for congestion, and twenty-seven cents for accidents. The small number for carbon dioxide emissions is surprising; it largely reflects the fact that there is less carbon in gasoline than one might think. Put another way, automobiles make a large contribution to global warming simply because of the sheer number of cars being driven and the amount of fuel consumed; on a per-gallon basis, the external costs due to global warming are fairly small. (The six cents/gallon estimate corresponds to marginal damages of twenty-five dollars per ton of carbon, which lies in the center of the range of estimates in the economics literature. Note that even doubling or quadrupling this number would still amount to a fairly small tax in cents per gallon terms.)

Parry and Small also point out that taxing gasoline is not necessarily the best possible policy. After all, the most costly externalities—local air pollution, congestion, and accidents—depend on the number of miles driven, rather than on the amount of gasoline consumed. Taxing gasoline, therefore, is just an (imperfect) proxy for taxing miles driven. (This is a bit like taxing steel when we really care about pollution.) The authors estimate that replacing the gasoline tax with a mileage tax would increase social welfare substantially. Their argument illustrates the general principle that the choice of *what* to regulate may be just as important (or more important) than the choice of *how* to regulate.

Filling in the Missing Demand Curve

We have just seen how a tax on output could restore the efficiency of markets, in the simple case where a unit of output always produces the same amount of pollution. But pollution and output are usually not so closely tied together. For example, an electric power plant can reduce its sulfur dioxide emissions by switching to a lower-sulfur coal or installing a scrubber, without reducing its electricity production. In this more general case, the efficient tax would be on pollution directly, not on output: that is, on sulfur dioxide, not on electricity. Looking at the problem this way will also lead to another intuition behind market-based instruments.

Figure 8.2 plots abatement (rather than output of a good like steel) on the horizontal axis. As in the graphs of marginal cost and benefit we drew in chapter 2, the efficient level of abatement is marked X^*, and coincides with the intersection of the marginal cost and marginal benefit curves. Looking at the figure now, however, you might notice something that may not have been apparent earlier: The marginal cost and benefit functions look an awful lot like supply and demand curves. Indeed, the resemblance is more than coincidental. Recall that the supply curve in a competitive industry is simply the marginal cost curve: Hence we can think of the marginal cost of abatement as the supply of abatement. Similarly, the marginal benefits from abatement can be thought of as representing the demand for pollution control.

Our discussions of market efficiency can shed new light on the efficient level of pollution control—and on how to design policies that will achieve it. Suppose that individuals and firms harmed by pollution could be induced to pay for pollution control according to their true valuation. In that case, there would be a demand curve for pollution control, tracing out the social marginal benefits from abatement. In that imaginary scenario, a market for pollution control would arise naturally—just as markets arise for other goods. The interactions of buyers and sellers would determine the price and quantity of pollution control, corresponding to the intersection of supply and demand. Assuming perfect competition and complete information, this outcome would be efficient.

The central problem, of course, is that such a demand curve never arises, because pollution control is a classic case of a public good. Every individual, comparing the cost of paying for pollution control (borne entirely by herself) with the benefit (shared by others), finds that it is in her self-

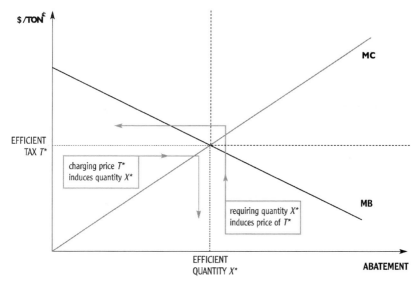

Figure 8.2   Market-based instruments to achieve efficient abatement. The efficient emissions tax equals the marginal benefits from abatement at the efficient quantity X^*. Note the equivalence between the emissions tax (a price instrument) and allowance trading (quantity instrument), in this case of certain marginal benefits and costs of abatement.

interest not to contribute. But when everyone free rides, the market demand for pollution control effectively falls to zero.[7] Note that the "supply curve" for abatement already exists: It simply corresponds to the marginal cost of controlling pollution. The hitch is that no firm will supply a good whose price is zero.

The role of government policy, then, can be understood as filling in the missing demand curve. Ideally, the government would reproduce the whole downward-sloping marginal benefit curve (yielding a downward-sloping demand curve). To do so, however, the government would need to know the entire marginal benefit function, and would have to be able to pay firms different amounts to reduce pollution depending on how much abatement was taking place.

Two simpler ways of filling in the missing demand curve are illustrated by the two dashed lines on the figure. First, the government can require a fixed quantity of pollution control. (See the vertical dashed line in the figure.) This corresponds to a tradable allowances program with a fixed cap on allowable pollution. In effect, a cap-and-trade policy is a commitment by the government to "buy" a fixed amount of abatement from firms in the regulated industry.

Alternatively, the government might set a fixed price for pollution control. (This is represented by the horizontal dashed line on the figure.) An abatement subsidy, for example, amounts to a promise by the government to buy pollution control from firms at a set price. Since (as we saw above) an emissions tax creates the same incentives as an abatement subsidy, we can think of a tax in similar terms. In effect, a tax on pollution amounts to charging firms for what they would pollute in the absence of regulation, and then paying them back for every ton they abate.

A cap-and-trade system and a pollution tax, therefore, are complementary ways of filling in the missing demand curve—one by setting a quantity, the other by setting a price. Whether the government completes the market with a vertical "demand curve" (a cap-and-trade policy) or a horizontal one (a tax), it chooses the curve that will intersect the "supply curve" at precisely the efficient level of abatement. As you can see from the figure, the tax that achieves this level of abatement is the price that corresponds to the intersection of marginal benefit and marginal cost. This should not be surprising: It is the same thing we saw in the simple example of the steel market in a previous section. The efficient tax is the marginal damage of pollution at the efficient outcome.

Under a cap-and-trade system, the government can pin down the level of abatement directly, by fixing an allowable level of pollution. What about the price of pollution? In the cap-and-trade system, the price of pollution equals the price of an allowance, which is determined by the market. From figure 8.2, you can see that this price equals the tax that would achieve the same level of pollution. That is, the price of an allowance under an efficient cap-and-trade policy will be exactly equal to the efficient tax! The tax and the cap-and-trade policy are essentially different ways of getting to the same point.

Creating Property Rights

As a final way of thinking about market-based policies, consider the problem of open-access resources that we discussed in chapter 5 and again in chapter 7. Unrestricted access to a resource typically results in overexploitation, as individuals act in their own self-interest rather than the common good. How can market-based policies help in this case?

A Tale of Two Trading Markets: Carbon Dioxide Emissions Trading

In the very first chapter, we outlined the possible consequences of global climate change resulting from the buildup of carbon dioxide (CO_2) and other greenhouse gases in the Earth's atmosphere. As attention to climate change has grown, so have the number of proposed approaches to limit CO_2 emissions. Given the great success of the sulfur dioxide trading program in the United States (which we will discuss in detail in chapter 10), emissions trading has become a leading framework for climate change reductions. Two CO_2 trading markets have been especially prominent: the European Union's Emissions Trading System (EU-ETS), and the Chicago Climate Exchange (CCX).[8] The contrast between the two programs is striking, and illustrates the important role played by government.

The EU-ETS is a government-run market, set up by the European Union (EU) to help meet its obligations under the Kyoto Protocol and modeled closely after the U.S. SO_2 program. Although the Kyoto obligations do not take effect until 2008, the EU established its carbon market in 2005 to smooth the transition. Even in its early phase (2005–2007), the sectors covered—energy, iron and steel, minerals manufacturing, and pulp and paper—face binding constraints on their CO_2 emissions. A cap was set for the EU as a whole, and allocated among member countries, which were then left with the task of dividing their allotments among affected polluting industries within their borders. The EU program is easily the largest one in the world. It covers roughly twelve thousand facilities in twenty-five countries, accounting for nearly half of the EU's total CO_2 emissions. The total allowance allocation is 2.1 billion tons per year, which implies a total market value of €30 to €40 billion ($37 to $50 million)—roughly ten times the total capitalization of the U.S. SO_2 market. In 2005, the first year of the program, 322 million metric tons of CO_2 equivalent were traded, representing a total value of $8.2 billion. After rising rapidly in the first half of 2005, prices have remained between €15 and €25 ($19 to $32) for much of the time since then.

CCX, on the other hand, is a purely voluntary exchange with a membership of twenty-four companies, municipalities, and universities, nearly all in the U.S. Members committed to reducing emissions by 4 percent by the end of 2006, relative to a baseline period of 1998–2001 (and, if they choose to continue, by 6 percent by the end of 2010). In 2005, three years into the exchange, just under half a million metric tons of CO_2 for that year's "vintage" exchanged hands, at a total value of roughly $850,000. Since the inception of the program in 2003, prices have risen from less than one dollar per ton of CO_2 to around four dollars as of August 2006.

A Tale of Two Trading Markets *continued*

Why have the fortunes of the two markets diverged so much? The answer, of course, is that the EU-ETS is a mandatory government program, while CCX is purely voluntary. As we saw in chapter 5, voluntary private provision of a public good is bound to be inefficiently low. Despite a great deal of hype surrounding CCX, it has had a miniscule impact relative to mandatory programs such as the EU-ETS. The contrast between the programs underscores the importance of the "cap" in "cap-and-trade." Without the regulatory power of government to enforce a binding cap on all sources within a region or industry sector, one cannot expect a voluntary program to accomplish much.

You have probably already guessed the answer. If we diagnose the problem as a lack of clear property rights, one promising remedy is to establish (and enforce) such property rights. Return for a moment to the shepherds in Hardin's parable of the tragedy of the commons. If the common pasture is divided among the shepherds, then each shepherd will have proper incentives to manage their own land wisely. Once private property rights are established, the market outcome will be efficient, since one shepherd's stocking decision no longer affects the productivity of pastureland for everyone else.[9]

How would this be applied to natural resource management in the real world? In the case of a fishery, a property-rights approach corresponds to what is known as *individual fishing quota* (IFQ) markets (discussed in more detail in the next chapter). Under such a policy, the total allowable harvest in a given year is divided up among a number of fishers. Each fisher receives an allotment or "quota," which confers the right to take a certain percentage of the total catch. Fishers can then buy and sell their quota on an open market, or lease them for a year. Such a policy is not simply a way of getting around open access, although it does achieve that goal. After all, traditional approaches to fishery regulation—for example, setting a limit on the total allowable catch, or restricting the length of the fishing season or the type of gear allowed—can also be seen as restricting entry or fishing effort, at least to some degree. The novel twist of the IFQ approach is that individual fishers receive *de facto* property rights in the resource, which they can trade among themselves. This gives the fishers themselves incentives not only to preserve the resource, but also to harvest the resource in the most efficient manner possible.

This property rights approach might also remind you of the Coase Theorem that we discussed in the beginning of the chapter. Indeed, the economist who popularized the idea of pollution trading (J. H. Dales, author of the 1968 book *Pollution, Property, and Prices*) made the connection explicit. Although pollution allowances are not private property in a legal sense, they do represent well-defined objects of trade whose value accrues to the holder. A market for tradeable allowances effectively converts a nonrival, nonexcludable public good (clean air) into a collection of private goods (allowances). Since the allowances are excludable and rival (they cannot be shared or held in common), they can be bought and sold in a market just like any other private good.

The more general point is that the crux of commons problems is the lack of exclusion. Thus we can extend the intuition of property rights to cases in which we don't literally divide up a resource among individuals. Consider the example we gave of highway traffic in chapter 5. Access to the highway can be restricted by erecting a toll booth—transforming an open-access resource into what is effectively a privately operated one (where the owner, in this case, is the government operator of the toll road). By raising the toll, the operator can reduce traffic and alleviate congestion. Indeed, the *efficient* toll is precisely equal to the external damages each additional driver imposes on everyone else—in other words, the Pigouvian tax! (This provides another example of the fundamental connections among our three ways of framing environmental market failures.)

In the case of managing a natural resource, the analog to charging a toll would be levying a "landing tax" on harvests. Just as a tollbooth excludes some people from using a highway, a landing tax on fish catches would in theory restrict the total harvest indirectly (through a price) rather than directly (through a system of property rights). However, while tradeable quota markets have been used to manage fisheries (as we will see in detail in chapter 10), landing taxes have not been implemented, presumably because of political opposition.

Is It Preferable to Set Prices or Quantities?

As we pointed out above, there is a deep underlying equivalence between an emissions tax and a cap-and-trade policy. One sets a price, the other a quantity; but as far as efficiency goes, the policies are just two ways of

Charging for Road Congestion in Singapore

As we have seen, road congestion is a prime example of the tragedy of the commons. The most costly externality from congestion is simply the time lost in traffic; but congestion also mean more pollution (since longer commutes mean more pollution) and a higher rate of traffic accidents. Economic theory suggests a simple prescription: Charge users a fee, or toll, to use the roadway.

Toll roads are commonplace in the U.S. and elsewhere. Charges aimed at reducing congestion have been implemented in many cities—most famously in London, where since 2003 drivers have been charged £9 (originally £5) to enter an eight-square-mile zone in the city center. But by far the world's most sophisticated system of congestion pricing can be found in Singapore, a city-state archipelago off the southern tip of the Malay peninsula in Southeast Asia.[10] With 4.5 million people occupying a land area of 270 km² (700 km²), Singapore is somewhat smaller in both area and size than New York City (population 8 million, area of just over 300 square miles).

In 1975, Singapore pioneered the implementation of a road-pricing scheme, known as the Area Licensing Scheme, which charged drivers sixty dollars per month for licenses to enter the city's central business district during peak hours. In September 1998, a new computerized real-time pricing system called "Electronic Road Pricing" (ERP) was introduced. Forty-eight overhead sensors, mounted on gantries located throughout the central business district and on major arteries, monitor traffic. Congestion fees are deducted automatically from prepaid CashCards inserted into In-vehicle Units at the start of a journey.

The goal of the program is to maintain average roadway speeds of 45 to 65 km/h (28 to 40 mph) on expressways and 20 to 30 km/h (12 to 18 mph) on city roads. To achieve this, the congestion charge varies by time of day and location. For example, a weekday trip by passenger car on the Central Expressway costs $1.50 between 7:30 and 8:00 in the morning, rises to $3 between 8:30 and 9:00, falls back to $1 between 9:00 and 9:30, and then is free until the next morning's rush hour. At the same time, charges on other (less heavily traveled) roads range from fifty cents to one dollar; the most expensive trip (on a stretch of the Pan-Island Expressway) costs $3.50. Charges are lower on weekends, and also differ for motorcycles, taxis, and various classes of trucks. The fee schedule is adjusted quarterly to maintain the target speeds.

Anecdotal evidence suggests a boost in car pooling and a shifting of traffic to off-peak hours. The system is credited with reducing peak-hour traffic by somewhat over 10 percent and increasing average road speeds by 20 percent. The evolution of the system also helps illustrate the logic of congestion problems. Installing a toll in one location diverts traffic elsewhere—that is, after all, part of the point. As smaller roads have gotten congested, the roadway authority has extended the system onto those roads, using the price differentials to influence traffic patterns.

getting to the same point. The price of an allowance (under an efficient cap-and-trade policy) will be exactly equal to the efficient emissions tax; and the pollution achieved by the tax will be the same as that imposed by the cap. This theoretical equivalence also applies at the level of the individual firm. Since the two policies create the same incentives for reducing pollution, the allocation of pollution abatement among the firms in the industry will also be identical under the two approaches. That is, the amount of abatement due to any particular firm will be the same under the tax as under a cap-and-trade policy.

This theoretical equivalence is worth emphasizing because it underscores the common intuition behind the two market-based policies. However, it turns out to rely on some hidden assumptions we have made: In particular, that the regulator knows the industry's marginal abatement costs, and that the environmental externality is the only distortion in the marketplace.

In the real world, marginal costs are often uncertain; and markets for labor, capital, and other goods are distorted by the taxes governments must levy to raise needed funds. It turns out that under these circumstances, the tax and cap-and-trade policies are no longer perfectly equivalent. The choice does matter for efficiency. The next question, of course, is: "Which is better?" The answer is: It depends.

Uncertainty about Marginal Costs

In drawing figure 8.2, we assumed that the marginal costs and benefits of abatement are perfectly known by the regulator. But what if the government lacks such precise information? It turns out that when marginal cost is uncertain, the choice of price versus quantity matters for efficiency. In particular, a tax (price) is preferable to a cap-and-trade policy (quantity) when the marginal benefit curve is flat relative to marginal cost, and vice versa. Surprisingly, however, uncertainty in the marginal benefit function does not matter for policy.

First, let's consider what happens when marginal cost is uncertain. Suppose that the regulator knows what the marginal cost curve will be on average, but not in any particular case at the time the policy is chosen. The actual cost could turn out to be above

In theory, an emissions tax and a cap-and-trade policy are just two ways of getting to the same point. In the real world, however, the choice does matter for efficiency.

or below this average or expected value. Figure 8.3 illustrates this case, depicting a high and low marginal cost curve, along with the average curve (labeled *EMC* for "expected marginal cost") and the marginal benefit curve, which we assume is known.[11]

This uncertainty can be understood in two ways. One interpretation is that marginal costs are unknown at the time that the regulation must be determined, and that the regulation is "sticky"—that is, it cannot be changed—over some period of time during which the marginal costs will become known. Alternatively, you can imagine that the regulated firms know their own marginal abatement costs all along, but for strategic reasons are unwilling to disclose them to the regulator.

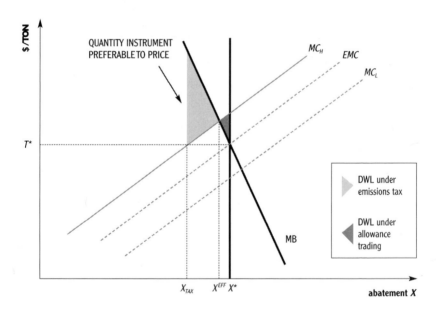

Figure 8.3 Comparison of price (emissions tax) and quantity (allowance trading) instruments under marginal cost uncertainty. The solid marginal cost line, denoted MC_H, represents the actual high marginal abatement cost curve, which is unknown to the regulator in advance. The bottom dashed line parallel to it, denoted MC_L, shows the alternative possibility (equally likely ahead of time) of a low marginal abatement cost curve. The middle line (*EMC*) is the expected marginal cost curve. The figure depicts a case in which marginal benefit is steeper than marginal cost, hence the cap-and-trade policy is preferable (smaller deadweight loss).

What should the regulator do? Let's consider a quantity instrument first—that is, a cap-and-trade policy. Since the regulator doesn't know the true marginal cost curve, she can't simply set the allowable pollution equal to its true efficient level. Instead, the best she can achieve is the outcome that is expected to be efficient—that is, the level of abatement that equates marginal benefits with expected marginal costs. We have labeled this $X^\star$ on the figure. Despite her best efforts, this will result in some inefficiency. Since the true marginal cost curve will be different from the expected value, it will intersect marginal benefits at some other level of abatement. On the figure, we have shown what would happen if marginal cost were higher than expected. The efficient level of abatement after the fact would be X^{EFF}, which is less than $X^\star$ (since firms find it more costly to reduce their pollution than the regulator anticipated). But of course the industry abates all the way up to the required amount, which is $X^\star$. The resulting deadweight loss is shown by the dark shaded triangle.

Now let's turn to a price instrument (i.e., a tax on emissions). The best the regulator can do is to set the tax $T^\star$ equal to marginal damages at the expected efficient level of abatement, $X^\star$. Again, suppose that marginal abatement cost turns out to be higher than expected, so that the efficient level of abatement is X^{EFF}. How will the polluting firms respond? They will abate until their marginal costs of pollution control are just equal to the tax. This level of abatement is labeled X^T on the figure. Beyond that point, the tax savings from more abatement are outweighed by the costs of pollution control. (We will see this in more detail in the next chapter.)

Under the tax, therefore, firms will do too little abatement when reducing pollution is more costly than expected. (If the regulator had better information on cost, she would impose a higher tax, which would be sufficient to achieve the efficient level of pollution control.) Again, some deadweight loss will be realized.

The key point is that the deadweight loss from allowance trading is not the same as the deadweight loss from the tax. The reason is that abatement differs in the two cases. Under a cap-and-trade policy, the amount of abatement is fixed at $X^\star$ by the cap. On the other hand, under a tax the amount of abatement varies with the true marginal cost.

Since the deadweight losses are different, it matters for efficiency which policy the regulator chooses. The preferred policy, naturally, is the one with the lower deadweight loss. In figure 8.3, you can see that the deadweight loss is smaller under the quantity instrument. But compare that to figure

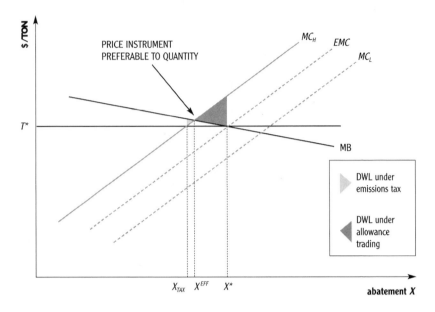

Figure 8.4 Comparison of price (emissions tax) and quantity (allowance trading) instruments under marginal cost uncertainty. In this case, marginal benefit is flat relative to marginal cost, and the emissions tax (the price instrument) is preferred. All else is as in figure 8.3.

8.4. In that case, the deadweight loss is smaller under the price instrument. What has changed?

If you look closely, you'll see that the difference in the two graphs is the *relative slopes* of the marginal benefit and marginal cost curves. In figure 8.3, the marginal benefit curve is steep relative to the marginal cost curve; in figure 8.4, the marginal benefit curve is relatively flat. In fact, those relative slopes are precisely what matters. This result is often referred to as the Weitzman Rule.

> When marginal costs are uncertain, the efficient choice of policy instrument depends on the slopes of the marginal benefit and marginal cost curves.

In particular, a price instrument is preferred when the marginal benefit curve is flatter than the marginal cost curve, and a quantity instrument is preferred when the reverse is true.

As we have seen, the two policy instruments differ in the amount of flexibility they give to the industry to respond to abatement costs. Uncertainty matters because firms can respond to the realization of marginal cost under the tax, but not under a cap-and-trade system. Whether this flexibility is a good thing or not (from the point of view of efficiency) depends on the relative slopes of the marginal benefit and cost curves.

A simple way to understand the intuition behind this relative-slopes rule is to focus separately on benefit and cost in turn. First, consider the slope of the marginal benefit function. Recall that we can think of the regulator's goal as replicating the missing demand for abatement—that is, the underlying marginal benefit curve. A tax represents a horizontal demand curve, which is a better approximation when marginal benefits are flat. When marginal benefits are steep, on the other hand, a vertical demand curve—corresponding to the quantity policy—is preferable.

You can also gain intuition by thinking about what the shape of the marginal benefit curve implies about the real world. Steep marginal benefits correspond to a threshold effect around $X^\star$, since the gains from further abatement drop off sharply at that point. (Conversely, the damages from pollution rise sharply as we increase pollution.) In this case, the regulator may want to ensure that the quantity target is met, since the damages from too little abatement will be very high. In contrast, a flat marginal benefit function implies no such urgency, since every ton of abatement brings roughly the same benefit.

Now consider the role of marginal costs. A flat marginal cost curve implies that the amount of abatement chosen by the regulated firms is highly sensitive to the tax, since a small change in marginal cost corresponds to a large change in abatement. Thus when marginal cost is flat, a small error in the size of the tax (and recall that some error is inevitable, given uncertainty) induces a large error in abatement. Loosely speaking, the potential costs to society of the abatement flexibility afforded by the tax are much greater when the marginal cost curve is flat. When the marginal cost curve is steep, the tax has much less effect on the actual level of abatement.

So far, we have only discussed uncertainty in marginal costs, rather than marginal benefits. Will the same arguments hold for the latter kind of uncertainty? The answer is "no." In fact, uncertainty in marginal benefits makes no difference in the choice of instrument. This might surprise you at first. But note that—unlike marginal costs—marginal benefits are irrelevant to the polluting firms' decisions. After all, if the firms took marginal benefits into account, there would be no externality in the first place! Since firms ignore marginal benefits in their own calculations, the fact that the regulator is uncertain about marginal benefits has no impact on which instrument it should choose. While deadweight loss will result (since the price or quantity chosen ahead of time will be different from the one that ends up being efficient), the actual abatement will be the same under both instruments, and thus the deadweight loss will be the same as well.

Raising Revenues

So far, we have ignored an obvious difference between taxes and cap-and-trade policies: Taxes raise revenue for the government, while cap-and-trade policies do not. Strictly speaking, the question of revenues does not depend on the use of a price or quantity instrument. We have already noted that pollution allowances can (in theory) be auctioned off, although in practice they have been given away for free. Meanwhile, an emissions tax need not be levied on all units of emissions (as we have implicitly assumed so far), but instead could be assessed relative to a very high baseline level of pollution.

Indeed, it is important to recognize that the amount of revenue raised by a tax or allowance auction is irrelevant to how well the policy deals with the negative externality. In correcting the market failure, what matters is that polluters have an incentive on the margin to reduce their emissions. They must face a price on the last unit of pollution. Whether they end up paying for every unit of pollution, or instead receive some pollution for free— as they do when pollution allowances are freely allocated—does not matter in terms of the incentives they face.

This might be surprising at first, but in fact it has been embedded in our argument all along. A firm's decision to abate 101 tons of pollution rather than 100 tons depends on the price of the 101st ton—not on the price of the first. You might think, "Well, if a firm is given one hundred pollution allowances for free, then it doesn't have any incentive to cut pollution below one hundred tons, since those emissions are all free." But now suppose that you are the manager of just such a firm, and suppose the market price of allowances is one hundred dollars. If you emit only ninety tons, rather than the one hundred you are allowed, you can sell ten permits and receive revenues of one thousand dollars. Indeed, each ton of pollution you emit incurs an opportunity cost equal to the price of an allowance, whether or not you buy the allowance directly or receive it for free.

Nonetheless, we can still ask: From the point of view of society, should the government raise revenue from environmental policies? And if so, how should it spend the money? Until recently, economists thought that these questions didn't matter, at least in terms of efficiency. After all, we have just seen that the revenues raised by a policy have no bearing on the abatement incentives created by the policy.

The key is to widen our perspective and take into account other sectors of the economy. Quite apart from environmental regulation, governments raise needed funds by taxing income, capital investments, corporate profits, consumer purchases, and so on. These taxes are typically distortionary, undermining the smooth operation of the market: For example, an income tax reduces the incentive to work, while a tax on capital affects investment decisions. Since income taxes hinder market efficiency, while environmental taxes (or cap-and-trade systems) increase it, a natural proposal is to use the revenue from emissions taxes (or allowance auctions) to fund reductions in distortionary taxes. That is, the government could cut the income tax, and make up for the lost revenues with an emissions tax.

This is sometimes referred to as a double dividend, since the pollution tax not only corrects the negative externality but also alleviates the distortions caused by taxes on income and capital.[12] Revenues do matter for efficiency, after all. From an economic standpoint, environmental policies should raise revenue from polluters, and use it to offset taxes on income, sales, and capital gains. Whether this would be feasible from a political point of view, of course, is an open question.

Conclusion

Now we've seen how economic theory can inform the design of environmental policy. We learned from Ronald Coase that there are conditions under which private bargaining will take care of negative externalities. In many cases of interest in the environmental realm, however, transactions costs will be large enough that government policies are needed.

In such cases, economics has much to say about what those policies should look like. Emissions taxes and cap-and-trade policies use market principles to restore the efficiency of the market. Whether we think of these policies as getting the prices right, as filling in for the missing demand for public goods, or as establishing property rights (and excludability) over common resources, the basic mechanism is the same: Market-based instruments align the incentives of private firms and individuals with the public interest. Finally, we saw how the choice between controlling price (through an emissions tax) or quantity (through a cap-and-trade system) can have important implications for efficiency.

So far, our discussion has been focused on how market-based instruments can achieve *efficiency*. In the next chapter, we'll see how a strong case for such policies can still be made, regardless of how the ultimate goal of the policy—for example, the level of abatement—is determined.

9

The Case for
Market-Based Instruments
in the Real World

Now we know how market-based instruments can be used—at least in theory—to restore the efficiency of markets. However, efficiency may not be
the relevant target in the real world, for several reasons. As we have seen, it
is very difficult to ascertain just how much benefit we get from these policies; indeed, it may even be hard to estimate the actual costs. But without
knowing costs and benefits, we cannot determine the efficient level of pollution. Even if the marginal benefits and costs of pollution control are
known, moreover, there is no guarantee that the government will set the efficient target as the goal. Distributional equity and other worthy social goals
may be at odds with efficiency. And of course the political process in the
real world is driven by interest group competition and the desires of legis-
lators to satisfy their constituents as much as (or more than) by an objective attempt to maximize social welfare.

Even if economic efficiency is elusive in practice, there are still considerable advantages to market-based instruments. First, such policies are cost-
effective, meaning that they can achieve any given abatement target at the
minimum total cost—something that is not generally true of command-
and-control approaches.[1] Second, over the long run, market-based instruments are likely to provide stronger incentives for the development of new
pollution-control technologies. This *dynamic incentive* will tend to lower
abatement costs over time.

In this chapter, we first discuss each of these advantages in detail for the
case of pollution control. Next, we consider how similar principles play out
in the realm of natural resource management, focusing on the case of a

fishery. Finally, we briefly review some of the other arguments for (and against) market-based instruments. Although they are powerful tools, market-based instruments are not panaceas. We close the chapter by discussing some conditions under which command-and-control approaches might be preferred.

Reducing Costs

In the previous chapter, we discussed taxes and cap-and-trade systems as ways of achieving efficiency. But the question of how much environmental protection to achieve can be separated from the question of how to achieve it. In other words, we can distinguish between goals (ends) and instruments (means). This distinction is key to understanding the notion of cost-effectiveness. Taking the policy goal as given, we may still ask: How do the various policy instruments perform in terms of the total cost of achieving that goal?

Imagine that you are a regulator who has been tasked with designing a policy to ensure that a specific pollution target is reached. For simplicity, suppose that there are only two polluters in the industry. In the absence of any regulation, each emits one hundred tons of pollution. You (the regulator) are supposed to cut the total in half—that is, you must achieve one hundred tons of abatement, so that *combined* emissions is only one hundred tons. The question is how to do that.

Figure 9.1 provides a useful way of depicting the problem. There are two firms—call them firm A and firm B. The horizontal axis measures abatement—but with a twist. Each point along the axis represents a different allocation of the hundred tons of total abatement between the two firms. As we move from left to right, the share of abatement done by firm A increases, while firm B's share decreases; their combined abatement stays constant. For example, at the left-hand corner of the figure, firm A's abatement is zero and firm B's abatement is one hundred tons. At the middle of the axis, each firm abates fifty tons. Finally, at the right-hand end, all one hundred tons of abatement are achieved by firm A, while firm B does nothing. The key thing to realize is that at every point along the horizontal axis, the total policy target (one hundred tons) is met. What varies is how that target is divided between the two firms.

The vertical axis measures marginal cost, in (say) dollars per ton. Consistent with the way we have measured abatement on the horizontal axis, firm A's marginal cost curve (labeled MC_A) increases from left to right, while

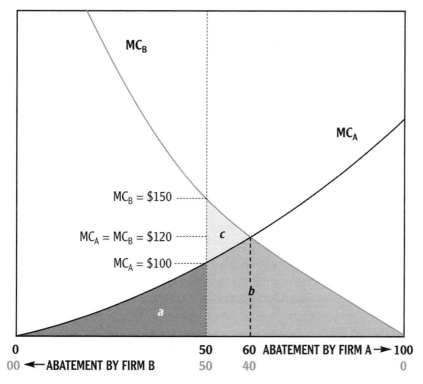

Figure 9.1 Marginal costs in a two-firm polluting industry. Note that the horizontal axis measures the allocation of a constant amount of abatement between firms A and B. The share of abatement done by firm A increases as one moves to the right. At every point, however, total abatement is one hundred tons. The shaded area labeled *a* represents Firm A's total abatement cost under a uniform standard. Areas *b* and *c* combined correspond to Firm B's total abatement cost under the same standard. The cost-effective allocation is where the two *MC* curves intersect. Area *c* represents the cost savings from the cost-effective allocation, relative to the uniform standard.

firm B's marginal cost curve increases from right to left. We have drawn the curves so that firm A is the "low-cost firm": that is, firm A can achieve any given amount of abatement at lower marginal cost than firm B.

Cost-Effectiveness

Remember that your task, as the regulator, is to find the policy instrument that minimizes the total abatement cost. A natural approach would be to split the abatement equally between the two firms, and require each firm to reduce pollution by fifty tons. The total abatement cost for firm A is the area under its marginal cost curve, labeled MC_A, up to fifty tons of abatement. On the figure, this corresponds to the shaded area labeled "a." Similarly, the total abatement cost for firm B is the area under MC_B up to fifty tons of abatement—the sum of areas "b" and "c" in the figure.

Is there a way to achieve the same total abatement at lower total cost? The figure suggests that there is. Starting from the uniform allocation, suppose that we increase the share of abatement done by firm A. That lowers total abatement cost: the sum of the areas under the marginal cost curves, up to each firm's level of abatement. Total abatement cost continues to shrink as we assign more abatement to firm A, as long as firm A's marginal cost curve lies below firm B's curve. In fact, the combined area under the two marginal cost curves is minimized where the two curves cross. Beyond that point, MC_A lies above MC_B. Continuing to assign greater pollution control to firm A, at that point, will increase rather than decrease total cost.

In this simple scenario, therefore, the cost-effective allocation of abatement—that is, the allocation that achieves one hundred tons of abatement at lowest total cost—is given by the point where the two marginal cost curves intersect. In figure 9.1, that corresponds to sixty tons of abatement by firm A and forty tons of abatement by firm B. The cost savings from this allocation—relative to the uniform standard—is represented by area "c."

It should not surprise you that the way to minimize total cost is to choose the allocation that equates the firms' marginal costs. Indeed, this is just another example of the equimarginal principle we saw in chapter 2. To see the intuition "on the margin," go back to the uniform standards case, when each firm must abate fifty tons. At that point, abatement is more costly on the margin for firm B than for firm A. (As we have drawn the figure, firm B's marginal cost at fifty tons of abatement is $150, while firm A's marginal cost is only $100.) Suppose that we took one unit of abatement away from firm B and assigned it to firm A. The amount of abatement would be unchanged, but the total cost would go down—by exactly the difference in the marginal costs, or fifty dollars.

The cost-effective allocation of abatement is the one that achieves a given level of pollution control at the lowest total cost.

As long as firm B's marginal cost is greater than firm A's marginal cost, we can continue to shift abatement from B to A—reducing cost without affecting abatement. This remains true until the two firms have equal marginal costs.

Note the important distinction between equal marginal abatement costs (given certain levels of abatement) and equal marginal abatement cost *functions*. We do not assume that firms have the same marginal abatement cost functions; in figure 9.1, for example, any given amount of abatement is more costly for firm B than for firm A. Moreover, each firm is doing different amounts of pollution control, with the low-cost firm A doing more abatement. Cost-effectiveness simply requires that abatement costs are equal on the margin, given the amounts of abatement that each firm is doing.

While we can only depict two firms in a graph like figure 9.1, the same intuition applies when there are many polluting firms. For example, consider a policy to achieve two thousand tons of abatement in an industry with one hundred firms. If that were divided equally among the firms, each firm would have to abate twenty tons. Now pick any two firms in the industry. If they have different marginal costs, we can shift abatement from the higher-cost firm to the lower-cost one until their marginal costs are equal. We can continue to do this with every possible pair of firms until it is no longer possible to shift abatement from one firm to another without increasing total costs. At that point, the marginal abatement cost is equal across all firms. In fact, we have just identified a key condition for cost-effectiveness:[2]

> Cost-effective allocation of abatement occurs only when all firms that abate pollution have equal marginal abatement costs, given their abatement allocations.

In plain English, the last unit of pollution control done by every firm must cost the same amount. Otherwise, there would be a way to reallocate abatement at lower cost.

Command-and-Control Approaches

Now let's consider how the various policy instruments we introduced in chapter 8 perform on this dimension of cost-effectiveness. First, consider technology standards, which require firms to install particular methods of controlling pollution. You may have already guessed from our discussion that such regulations are not cost-effective in general. In terms of our "necessary condition," technology standards typically fail because different firms will have different costs of installing the same sort of technology. For example, some power plants may have plenty of space to build a scrubber, while other plants will have to build specially designed units to fit into a limited footprint on the ground. More fundamentally, technology standards are generally not cost-effective because they do not even minimize costs at the level of an individual polluter! Some firms are likely to have other ways of limiting pollution that could achieve the same low levels of emissions as the required technology, at lower cost. Another way to reduce sulfur dioxide emissions from power plants is to burn very low-sulfur coal. Modifying a plant to burn low-sulfur coal, and even paying a premium for it (versus cheaper high-sulfur coal), is often much less expensive than installing and operating a scrubber—but nearly as effective in reducing pollution. We will see a vivid example of how costly the lack of flexibility can be in the real world, when we consider the case of sulfur dioxide control in chapter 10.

Next, let's consider performance standards, which impose emissions limits on individual firms. Return to our two-firm example, with a total emissions target of one hundred tons. A uniform performance standard would set a ceiling of fifty tons for each firm. Since we have assumed that firms A and B would emit the same amount of pollution (one hundred tons each) in the absence of regulation, this uniform standard on emissions amounts to a requirement that each firm does the same amount of abatement. We have already seen that such uniform regulation is not cost-effective in the case of figure 9.1. Indeed, from our discussion of cost-effectiveness we can now see that uniform standards will never be cost-effective, as long as firms have different marginal abatement cost functions—that is, as long as firms face different opportunities to reduce their emissions.

"But wait a minute," you might say, "that is a problem with *uniform* standards—not with performance standards in general!" Indeed, that is correct. If you (as the regulator) knew the marginal cost curves of the two firms, you could establish firm-specific performance standards corresponding to the cost-effective allocation. In the scenario of figure 9.1, this would mean setting emissions standards of forty tons for firm A (requiring sixty tons of abatement) and sixty-five tons for firm B (forty tons of abatement).

While cost-effective performance standards are theoretically possible, however, they impose an unrealistically high informational burden on the regulator. Firms have obvious incentives to misrepresent their true marginal cost curves: In our simple two-firm example, each firm would like the regulator think that it was the high-cost firm. Deriving firm-level marginal cost functions without the firms' cooperation, meanwhile, would require data far more detailed than what real-world regulators have. As we shall now see, market-based instruments can achieve cost-effective allocations even when the regulator has much less information about abatement costs.

Emissions Taxes

To see why an emissions tax is cost-effective, let's consider what an individual firm will do when faced with a tax. Figure 9.2 depicts firm A's marginal cost curve by itself. As usual, the horizontal axis measures pollution abatement. We have denoted firm A's maximum abatement (unregulated emissions) on the axis by X^{MAX}.

The dashed line on the figure represents the emissions tax. At any given level of abatement, firm A's total compliance cost is the sum of its abatement cost (the area under the marginal cost curve) and its tax bill (the area of the rectangle under the dashed line and to the right of the abatement level—that is, the tax times emissions, where emissions are just the difference between abatement and unregulated emissions). This compliance cost is minimized at the point where marginal abatement cost equals the tax (de-

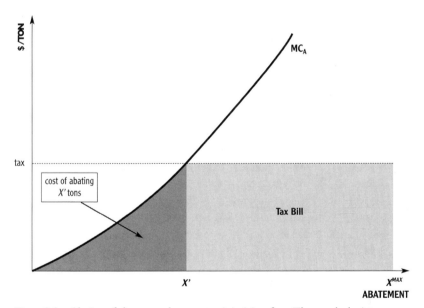

Figure 9.2 Choice of abatement by a cost-minimizing firm. The graph depicts a marginal abatement cost function for an individual firm. The cost of compliance is the sum of the tax bill and the cost of abatement—the two areas shaded in the figure. The least-cost abatement level X' is the level at which marginal abatement cost equals the tax.

noted X' on the figure). (This is just like a cost-minimizing firm that produces up to the point that its marginal cost of production equals the price it receives for its output.) If the firm were to abate less than X', its compliance costs would be higher, because it would pay more in emissions taxes than it saved in abatement costs. On the other hand, costs of more abatement would exceed the tax on the margin.

Nothing is special about firm A in this regard. Indeed, we could repeat the same analysis with firm B, or indeed with any firm. In every case, a cost-minimizing firm would choose the abatement level to equate its marginal abatement cost with the emissions tax.

What does this imply about cost-effectiveness? Since every firm is setting its marginal abatement cost equal to the tax, and every firm faces the same emissions tax, it follows that marginal abatement cost is equal across all the polluting firms. Therefore, the condition for cost-effectiveness is met. Return to figure 9.1, which depicts firms A and B on the same set of axes. We showed above that the cost-effective allocation (sixty tons by firm A, forty tons by firm B) coincides with the intersection of the marginal abatement cost curves. Notice also that the cost-effective allocation is the only point at which the marginal abatement costs of the two firms are equal. Since the emissions tax ensures that the two firms have the same marginal abatement cost, it follows that the emissions tax achieves the cost-effective allocation. Again, the cost savings are represented in the figure by the shaded triangle labeled "c."

Moreover, all this happens without direct intervention from the regulator! All the regulator does is to set the tax. Given that tax, each firm independently chooses to abate at the level such that its marginal abatement cost equals the tax. In doing so, it ensures that its marginal abatement cost is equal to that of every other regulated firm.

Of course, for the regulator to successfully implement the tax, she must know how high to set the tax in order to achieve the desired policy target. In particular, the regulator must know the *aggregate*, or industry-level, marginal abatement cost curve. Such a curve traces out the cost of the last unit of abatement, as a function of the total abatement done by the industry as a whole. It is derived by summing up the abatement done by all the firms in the industry, for any given marginal cost. Figure 9.3 provides an illustration. In it, we have drawn the marginal abatement cost curves corresponding to firms A and B depicted in figure 9.1, along with the aggregate marginal abatement cost curve corresponding to the two firms put together.

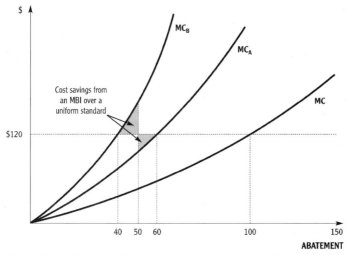

Figure 9.3 Individual and aggregate marginal cost curves. The aggregate curve (denoted *MC*) is the sum of the quantities abated by the two firms, at every level of marginal cost.

We have already noted the parallel with a supply curve in a market for private goods. In particular, one can think of the aggregate marginal abatement cost curve as the "abatement supply curve."[3] Recall that in the standard market setting, the supply curve tells us how much an industry will produce at a given price—or equivalently, how high the price of a good has to be to induce firms to produce a given quantity. In exactly the same way, the aggregate MC curve summarizes the relationship between the marginal cost of abatement and the total amount of abatement. Since the tax acts like a price on abatement, the total abatement induced by a tax can be found from the point on the aggregate MC curve corresponding to the tax. In the case of our two firms, A and B, the required tax is $120. That number corresponds to the height of the aggregate MC curve at one hundred tons of abatement.

As in figure 9.1, we can depict the cost savings from using the emissions tax (rather than a uniform standard) directly on the graph. In this case, the cost savings are represented by the two shaded areas shown. Close inspection should convince you that these two areas together correspond precisely to the shaded triangle that corresponds to cost savings in the earlier figure.

Tradable Allowances

Finally, let's consider a cap-and-trade policy. We can start by returning to figure 9.1. Once again, imagine that you are the regulator charged with limiting the combined emissions of firms A and B to one hundred tons. Now consider the following policy. Suppose you start by assigning each firm fifty tons of emissions, as in the case of uniform performance standards. But instead of requiring the firms to meet that standard individually, you give each firm fifty pollution allowances (worth one ton each) and allow them to trade. What will happen?

At the initial allocation of fifty tons of emissions each, each firm must abate fifty tons. At that level of abatement, firm A has a marginal cost of $100, versus $150 for firm B. Therefore, a trade will be in the interests of both firms. In particular, the manager of firm A could contact his counterpart at firm B and propose to sell her one allowance. Firm A would reduce its emissions by one ton, and sell the resulting extra pollution permit to firm B. Firm A would be willing to do this for any price above $100, while firm B would be willing to pay up to $150. In the language of economics, there are "gains from trade." As long as the two firms can easily trade with one another, therefore, we would expect them to make such a deal.

These gains from trade remain as long as the marginal costs of the two firms are not equal. In other words, the two firms will have an incentive to keep trading until their marginal costs are equated. But this means that trading achieves the cost-effective allocation of abatement! In terms of figure 9.1, we would expect firm A to sell ten pollution allowances to firm B. After trading, firm A would end up abating sixty tons (and receiving some payment from firm B), while firm B would abate only forty tons (but would pay firm A for the ten tons it was not abating). In terms of emissions, firm A would emit forty tons (below its initial allocation of fifty permits) while firm B would emit sixty tons (ten more than its allocation).

Moreover, note that this argument does not depend on how the allowances are initially allocated. Suppose, for example, that firm A receives only twenty allowances, while firm B receives eighty. (Total allowable pollution is one hundred tons, just as it was before.) Without trading, firm A would therefore be required to abate eighty tons, while firm B would have to abate only twenty tons. Once again, there are gains from trade. This time, however, the marginal cost of the last unit of abatement is higher for firm

A than for firm B. In figure 9.1, firm A's marginal cost curve is above firm B's marginal cost curve when firm A abates eighty tons and B abates only twenty tons. Given this initial allocation, therefore, firm A would buy allowances from firm B. The gains from trade would remain until the marginal costs were equal, at the cost-effective allocation of sixty tons of abatement by firm A and forty tons by firm B. We conclude that as long as firms can easily trade with one another, the initial allocation of pollution allowances does not affect the final (equilibrium) allocation.[4]

We have described this cap-and-trade system in the context of only two firms, but the logic generalizes immediately to any number of firms. As long as all firms take the price of pollution allowances as given—that is, as long as no firm has the power to set prices in the allowance market—then each firm will abate up to the point at which its marginal abatement cost equals the allowance price. Below that point, reducing pollution is less expensive than paying for permits. Beyond that point, it would make more sense for the firm to buy allowances rather than abating. Of course, since the total number of allowances is capped by the regulator, the number of allowances that are bought by firms who choose to abate less than their initial allocations must be exactly balanced by the allowances that are sold by firms who abate more than required.

Since every firm abates until its marginal cost equals the allowance price, and all firms face the same allowance price, marginal cost must be equated across all firms. Just like an emissions tax, therefore, a cap-and-trade policy is cost-effective. The market mechanism ensures that abatement is carried out by the firms who can do it at least cost.

One potential advantage of cap-and-trade systems is worth pointing out. The goal of environmental policy regulation is often defined in terms of the quantity of pollution (or of environmental protection more generally). If so, a cap-and-trade system is easier to implement than a tax. This is because the connection between an emissions tax and the resulting quantity of emission is indirect: the regulator sets the price, while the actual level of pollution depends on the industry's marginal costs of abatement. Therefore, the regulator must know the aggregate MAC curve in order to attain a particular level of abatement with an emissions tax. In contrast, a cap-and-trade system determines the quantity of pollution directly; no other information than the policy target is required.

Cost Savings and the Degree of Heterogeneity

You may have realized by now that the cost-effectiveness advantage of market-based instruments depends on the variation (or "heterogeneity") in the marginal abatement cost functions of different firms. To see this graphically, return to figure 9.1, and imagine redrawing the figure with firm A's marginal abatement cost function somewhat flatter than it is, and firm B's marginal cost function even steeper. Doing so would increase the cost savings from a cap-and-trade program or emissions tax, relative to a uniform standard. (Recall that those cost savings are represented by the shaded area labeled c in the figure.) On the other hand, if firms were identical, the cost-effective allocation of abatement would coincide with a uniform standard, and there would be no cost savings from a market-based instrument. The intuition is simple: The greater the differences between the firms, the greater will be the gains from allowing them to trade among themselves.

Promoting Technological Change

The notion of cost-effectiveness we have been using is essentially a static one. If a regulator has a certain target in mind for pollution control, for example, market-based instruments can achieve that goal at lower cost than would be possible through uniform performance standards. They do this by reallocating pollution control from firms with high costs of abatement to firms with low costs—although such reallocation results not from any centralized coordination, but from the actions of individual firms pursuing their own profits.

Our description of how market-based instruments work has essentially relied on a snapshot of the polluting industry at a particular point in time. For example, in the previous section we just asserted that firm A has lower marginal abatement costs than firm B—but we did not ask where that cost differential comes from. We simply took the marginal abatement cost functions for granted. Nor did we ask whether firm B might be able to upgrade its pollution control technology—say, by investing in new equipment, thereby lowering its marginal abatement cost. In other words,

The cost-effectiveness advantage of market-based instruments depends on the variation (or "heterogeneity") in the marginal abatement cost functions of different firms.

we implicitly viewed abatement technology as fixed.

Cost-effectiveness is an important criterion, but it is limited by the static perspective we have just described. A complementary means of assessing policy instruments is to ask how well they promote technological change— for example, the development and

The incentive to adopt new technologies with lower marginal costs is greater under an emissions tax than under a performance standard.

adoption of new pollution control technologies. It turns out that market-based instruments outperform standards on this dimension as well. More precisely, the incentive to adopt new technologies with lower marginal costs is greater under an emissions tax than under a performance standard. The incentive under a cap-and-trade system is less than under the tax, but is still likely to be larger than under the performance standard.

To see this, let's start by comparing an emissions tax with a performance standard. Figure 9.4 depicts a firm whose initial marginal abatement cost function is denoted MC_0. Another technology becomes available, with marginal cost MC_1. Is the firm more likely to adopt the new method under the tax, or under the performance standard? The gain to the firm from adopting the technology depends on the cost savings to the firm from having marginal cost function MC_1 rather than MC_0. The greater these cost savings, the more likely the firm is to adopt the new technology.[5]

We can use figure 9.4 to figure out how the reward from adoption varies under the two policies. To focus narrowly on the question of technology adoption, let's assume that the tax and standard are *equivalent* under the initial cost function MC_0. In other words, the firm would choose exactly the same amount of abatement under the tax as it would be required to do under the performance standard. On the graph, this means that the tax (represented by the horizontal dashed line) intersects the initial marginal cost curve at the emissions standard (the vertical dashed line at X_{STD}).

Note that under the performance standard the firm has no incentive to abate more than it is required to. Thus even if the firm adopts the lower-cost technology in that case, it will continue to abate the amount X_{STD}. The cost savings from the new technology (not counting the capital cost of installing the equipment, which we have not specified) correspond to the area of the shaded triangle labeled *a* in the figure.

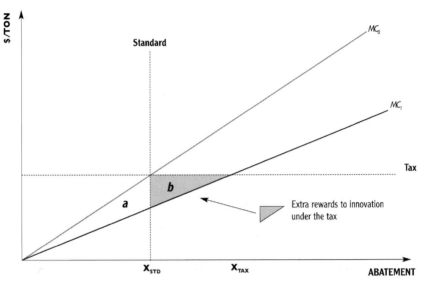

Figure 9.4 The incentive to adopt a new pollution control technology under a performance standard and an emissions tax. The two policies are assumed equivalent under the orginal technology (MC_0). The new technology MC_1 has fewer costs; its adoption results in greater abatement under the tax, and hence greater cost savings to the firm.

What is the reward from adoption if the firm is regulated by the tax instead? Under the tax, the firm abates up to the point that marginal cost equals the tax. Therefore, with the new technology the firm would abate all the way to X_{TAX}, which is greater than X_{STD}. Intuitively, a tax provides an incentive to abate as much pollution as is economically sensible. With a lower-cost abatement technology, it makes sense for the firm to abate more pollution, since it has to pay taxes on whatever it continues to emit. Therefore, the increased abatement translates into additional cost savings. This is represented by the area of triangle *b* in the figure.

Intuitively, we can think of area *a* as "cost savings from doing what the firm was already doing under the old technology, only cheaper." We can think of area *b* as "cost savings from increasing abatement, now that it's cheaper, rather than paying emissions taxes." The firm gets area *a* from adoption under both the standard and the tax; but area *b* is relevant only under the tax. We can conclude that the tax creates a greater incentive for the adoption of new abatement technologies.

A cap-and-trade system does not lend itself as well to a comparison like the one in figure 9.4, because the "cap" applies to the market as a whole, not to individual firms, and because different firms will necessarily have different cost functions (if they did not, there would be no gains from trade!) Nonetheless, we can glean some intuition by comparing a cap-and-trade system to a tax. As we have noted several times, a tax fixes the price of abatement, while a cap-and-trade system fixes the quantity. This distinction has an important implication for technology adoption. As more and more firms in an industry adopt the new technology, the price of tradeable allowances will fall, since abatement becomes less expensive on the margin. This fall in the price, however, *reduces* the incentive for other firms to adopt the technology. (You can see this on figure 9.4 by shifting the tax line down, representing a fall in the price of pollution.) This should make sense: After all, if allowances become less expensive, firms will have less reason to look for a cheaper way of reducing pollution.

This feedback between technology adoption and the price of pollution means that cap-and-trade policies will typically provide less incentive than a tax for the adoption of new abatement technologies. (This will be true if the tax and cap-and-trade system achieve the same level of abatement when firms have the original abatement technology MC_0, since the allowance price will fall while the tax remains fixed.)

Comparing a cap-and-trade system and a performance standard, meanwhile, turns out to be less straightforward. But the basic advantage of market-based policies continues to hold. Since firms pay for each unit of pollution they emit, they have a strong incentive to find ways to lower their marginal abatement costs. In general, then, we expect technology adoption to be greater under either an emissions tax or a cap-and-trade policy relative to command-and-control.[6]

You may have noticed that so far we have concentrated on performance standards, rather than on technology standards. Would a technology standard do better promoting technological development? At first, you might think that it would have to. After all, under such a policy all firms have to adopt the required technology. Doesn't that have to mean that technology standards provide the greatest adoption incentive of all?

Two distinctions are key here. The first is between adopting a particular technology mandated by the government, and adopting a lower cost technology. While we have described the gains from technology adoption in terms of just one new technology, the intuition applies even when there are many ways of reducing pollution. Faced with a menu of options, each firm will choose the one that suits it best—providing the greatest abatement cost savings in return for the investment required. On the other hand, a technology standard requires firms to choose a particular technology, regardless of its other options. Moreover, there is no reason to think that the government will have better information than firms about the true costs of installing and operating various technologies.

The second key distinction is between adoption and innovation. Our discussion so far has essentially assumed that lower-cost technologies simply pop into existence, awaiting firms to adopt them. In fact, that ignores the crucial role of innovation: the development of new technologies by firms that hope to sell them. Because market-based instruments give polluters a continuing incentive to search for new and better ways of reducing pollution, they provide a constant spur to innovation. On the other hand, technology standards run the risk of locking in a particular technology—whatever is the state-of-the-art at the time of the regulation—and thereby unintentionally dampening the incentives for firms to come up with new technologies.

Marked-Based Instruments for Managing Natural Resources

In chapter 7, we showed how open-access conditions lead to market failure in fisheries and other natural resources. As we saw there and in chapter 8, efficient management of a fishery could be achieved by establishing property rights, either in fact or in effect. But just as in the case of pollution control, efficient management of a fishery (or other resource) requires a great deal of information—not only the marginal costs of harvesting, but also the growth function of the resource. In the first part of this chapter, we saw how emissions taxes and allowance trading can lower the cost of meeting a given target for pollution control—whether or not that policy goal is efficient. Similarly, we saw how such market-based policies could encourage the adoption and development of new technologies over time.

Now we look at how market-based instruments can be used to manage natural resources. We focus on fisheries management as our example, but as always, the lessons are general. In chapter 10, for example, we see how these same principles apply to resources as diverse as scarce water supplies and valued wetland ecosystems.

Cost-Effectiveness

The key intuition is that cost-effectiveness means the least-cost allocation of effort to achieve a particular goal. In the case of a fishery, the goal is defined in terms of the annual allowable harvest, while effort means the time and money fishers spend on catching fish and purchasing and maintaining their equipment.

In chapter 7, we assumed for simplicity that the cost per unit of fishing effort—that is, the marginal cost of fishing—was constant (at least for a given size of the fish stock). In effect, this means that all fishers are identical, and that the last fish they catch requires no more effort than the first. In reality, of course, fishers differ widely in ability and cost. Some individual fishers are better able to read the ocean floor, or the patterns of seabirds and other animals, in order to locate stocks of fish. Large trawlers have lower costs (per unit of fish caught) than small one- or two-person boats. Moreover, it is reasonable to suppose that for any given fisher, the marginal costs of fishing increase with the total catch. At some point, for example, a fisher wishing to catch more fish must venture farther out to sea, or to less promising (or less familiar) fishing grounds.

Why does such variation matter? Previously we saw how the degree of heterogeneity among marginal abatement costs determined the cost savings from market-based policies for pollution control. The principle is exactly the same in the case of a fishery: the more varied the costs of fishing effort, the greater the savings from a cost-effective allocation.

To see how this works in the case of a fishery, consider figures 9.5 and 9.6. Suppose there are two fishers in a fishery, who face the same price for fish (which we will take to be fixed) but have different marginal costs of fishing. These assumptions are depicted in figure 9.5, where we have drawn two marginal cost-of-fishing curves, labeled MC_A and MC_B in the figure, and a horizontal line representing the market price of fish. As the figure is drawn, fisher B is the low-marginal-cost fisher. Suppose that there were no restrictions on fishing. In that case, each fisher would earn marginal net benefits (equivalently, marginal net revenues) equal to the difference between the market price (the revenue on each unit of fish) and the marginal cost. In plain English, the marginal net benefit represents the value to a fisher of each fish she catches. Since marginal costs rise while the price stays constant, marginal net benefits to each fisher fall as her catch increases. Moreover, since fisher B has lower marginal costs of fishing, she earns greater marginal net benefits.

Now take a look at figure 9.6. (This figure recalls both the two-firm pollution abatement model of figure 9.1 and the two-period resource extraction model of figure 6.3.) The two fishers are represented in the figure by

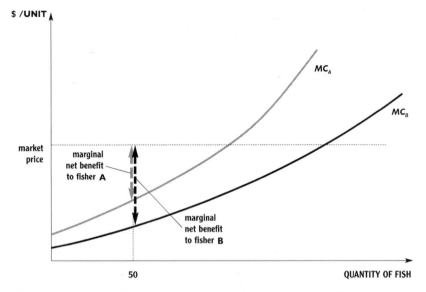

Figure 9.5 Marginal fishing costs, the market price, and the resulting "marginal net benefits" for two fishers in a market. When each fisher catches fifty fish (the line shown on the graph), the shorter arrow represents the marginal net benefit to fisher A, while the longer arrow represents the marginal net benefit to fisher B, who has lower marginal costs.

their marginal net benefit curves, labeled MNB_A and MNB_B. The length of the horizontal axis represents the total allowable catch in the fishery—say, one hundred tons of fish. The question is how to allocate that total catch between the two fishers. Suppose we start by dividing it equally. In this case, each fisher would harvest fifty tons. At that point, however, fisher B values the last ton of fish she catches more highly than fisher A values the last ton of her own harvest. (We know this because when each fisher lands fifty tons, the marginal net benefit for B is greater than the marginal net benefit for A.)

Now consider what would happen if instead of allocating the catch equally, we gave each fisher an equal number of individual fishing quotas (IFQs) and allowed them to trade. What will happen? Since the low-cost fisher B values a ton of fish more than her counterpart (on the margin), she will buy IFQs from fisher A. Indeed, as long as transactions costs are negligible, we would expect the two fishers to trade until their marginal net benefits are equated. This point corresponds to the intersection of the two

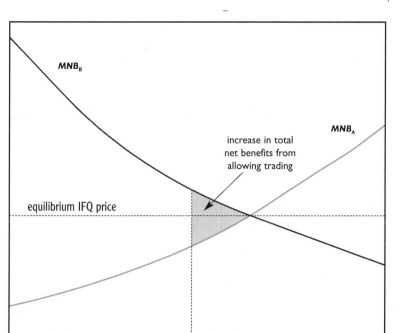

Figure 9.6 Marginal net benefits in a two-person fishery. The horizontal axis measures the total allowable catch, allocated between the two fishers. Note that the share of the harvest caught by fisher B increases as one moves to the *right*. At every point, however, the total catch is one hundred tons. The cost-effective allocation is where the two *MNB* curves intersect; the corresponding IFQ price is denoted by the horizontal dotted line. The shaded triangle represents the gains from an IFQ relative to a even allocation of the harvest. Under an IFQ, the low-cost fisher (B) catches more, while the high-cost fisher (A) catches correspondingly less. Since the marginal net benefits are higher for B, this represents a gain from the point of view of society.

MNB curves. The horizontal dotted line on the figure shows the price at which the last quota would trade. Moreover, the fishers will reach this point regardless of the initial allocation of the IFQs—just as we saw in the case of a pollution market.

The shaded triangle in figure 9.6 represents the total gains from trade, relative to the scenario in which each fisher lands half of the total catch. This is analogous to the cost savings from trading that we found in the case of pollution control (compare it to area c in figure 9.1). However, the triangle has a slightly different interpretation in this case. Instead of minimizing total costs, we are now interested in maximizing total net benefits. The net benefits to each fisher are the areas under their respective MNB curves. The gains from an IFQ system result because the low-cost fisher (fisher B in the figure) ends up with a larger share of the catch. Her lower costs of fishing translate into higher net benefits. The resulting increase in net benefits to the two fishers combined is shown by the shaded triangle.[7]

Incentives for Technological Change

In the case of pollution control, we saw how an emissions tax (and to a lesser extent a cap-and-trade policy) could encourage the adoption and innovation of new, low-cost abatement technologies. What analogous incentives are created by market-based policies for managing a natural resource? It turns out that less theoretical work has been done on this point, but we can sketch a few basic principles. First, an IFQ market (or a similar policy in another realm of natural resources) establishes secure property rights over the resource. In the absence of such property rights, as we saw in chapter 7, each individual user has a strong incentive to ignore the effects of her behavior today on the state of the resource tomorrow. The first-order dynamic incentive created by securing property rights, therefore, is to give each participant a stake in the continuing value of the resource—encouraging them to act in ways that sustain the resource's productivity.

Second, an IFQ market changes investment incentives. As we saw in chapter 7, under open-access conditions fishing effort will increase until the rents from the resource are exhausted. In the simple analysis of that earlier chapter, we measured effort in terms of the number of boats. More generally, however, we might think of effort as also encompassing the capital expenditures fishers make in their boats. For example, if a limit is placed on the total allowable catch, without restrictions on allowable gear, each fisher will have a strong incentive to buy the biggest, fastest boat possible, in order to take as many fish as she can before the total allowable catch runs out for the season. Attempts to restrict the catch by shortening the fishing season create similar incentives. The result is overcapitalization: Fishers spend much more on equipment than is needed to catch the harvest in an arms race with each other. In economic terms, each fisher sees only an incentive to increase her own share of the rent; in striving to do so, fishers as a group drive the rents to zero.

The security created by an IFQ market transforms these incentives. A fisher who is guaranteed a certain percentage of the catch has the proper incentive to invest enough to minimize her costs of harvesting that amount—but no more.

Other Considerations

Cost-effectiveness and the promotion of new technologies represent the cornerstones of the economic argument for market-based instruments. Choosing the best policy in a particular case may involve a number of other considerations, however. While a full treatment is beyond the scope of this book, we highlight a few of the main ideas in this section.

Pollution "Hot Spots"

In our analysis so far, we have implicitly assumed that "all emissions are created equal"—in other words, that the social damages from emissions are the same across all firms. This is a convenient assumption in modeling, because it means we do not have to keep track of who is emitting what. For example, in the model of the first section, we assumed that the relevant measure of pollution was the sum total of emissions from firms A and B combined. Economists call this the "uniform mixing assumption." If emissions from different sources uniformly mix in the atmosphere, then from society's point of view all that matters is total emissions, rather than the identity of the emitters.

Is this a reasonable assumption for the real world? The answer—perhaps not surprisingly—is, "It depends." In some cases, uniform mixing is a good description of the real world. For example, greenhouse gases such as carbon dioxide are believed to disperse widely in the atmosphere, regardless of where they come from. Thus a ton of CO_2 emitted in Boston and a ton in Beijing have the same effect on atmospheric concentrations of CO_2 (and hence on global warming). In other cases, however, pollution accumulates locally near where it is emitted, rather than dispersing widely. As an extreme example, consider toxic waste, which poisons soil at a specific site, and may seep into groundwater nearby, but does not have effects outside a well-defined watershed.

The degree of mixing matters because the policies we have described rely on uniform mixing to work properly. A cap-and-trade approach is sound policy only if a cap on total emissions makes sense, and if allowing firms to trade among themselves will not lead to dangerously high concentrations of pollution in localized areas. The same caveat applies to an emissions tax, which treats all sources the same (a ton of pollution is taxed the same regardless of its origin) and shifts pollution among firms just as a cap-and-trade does. A nationwide cap-and-trade system (or tax) limiting the aggregate quantity of toxic waste would be inappropriate, since it would ignore the question of how highly concentrated the waste was in any particular site. This is known as the hot spot problem: when pollution is nonuniformly mixed, cap-and-trade programs and emissions taxes can result in hot spots of concentrated pollution. Under a trading program or a tax, there is nothing to prevent emissions from individual sources to increase, even as overall emissions fall.

This issue of nonuniform mixing turns out to be particularly significant for water pollution. In most cases, the location of an effluent source strongly affects pollution concentrations at various receptor points throughout a water body. Marginal damages from pollution can also vary with seasonal or daily weather conditions. While source-specific trading ratios or effluent fees are possible in theory, they have proven difficult to implement in practice.

Even if hot spots don't arise, simple allowance trading programs that focus only on total pollution may lead to wide gaps in the damages due to pollution from different areas. What should we do in such situations? One solution, of course, is to fall back on command-and-control approaches, such as performance standards. In some cases—for example, toxic effluents released by paper mills into streams—this may prove to be the most sensible policy. Another solution might be to prohibit some kinds of trades. For example, we might prevent polluters in a high-damage area from buying allowances from firms in a low-damage area. This has its own problems, however. For one thing, it is unlikely to make much difference, as long as the market can circumvent the restriction.

Let's take an important water quality problem as an example.[8] In the Upper Ohio River Basin, there are approximately seventy sewerage systems, including that of the city of Pittsburgh and many smaller cities, with combined sewer overflows. These sewerage systems receive surface water runoff during rainfall, and when the flow exceeds the system capacity, raw sewage is discharged directly into waterways in the basin. The impacts of this include pollution damages from bacteria, biological oxygen demand (BOD), and total suspended solids. Suppose we set up a cap-and-trade system in which the cap was a limit on total allowable sewer overflows. The marginal damages from sewer overflows vary greatly, depending on the receiving water characteristics, including flow and other hydrological aspects of the water's capacity to assimilate waste, as well as the exposed population.

When pollution is nonuniformly mixed, cap-and-trade programs and emissions taxes can result in hot spots of concentrated pollution

To avoid hot spots, we could simply disallow trades between selected pairs of the seventy discharging sewerage systems, but this would be difficult to enforce. For example, suppose we forbid Pittsburgh from selling permits to another city, like Morgantown, because the marginal damages from Morgantown's emissions are higher than Pittsburgh's. In this case Pittsburgh could still sell permits to the sixty-eight other cities in the basin, and some of those permits could then be purchased by Morgantown in separate transactions. All that would have changed is the transaction cost. And if the restriction did limit trade, it would effectively reduce the gains from using a market-based instrument in the first place; the more divided the market, the smaller the potential gains from trade.

However, there is an interesting market-based alternative (although one that has not yet been implemented in practice). Conceptually, we might want to make it more expensive for Morgantown to buy permits from Pittsburgh. If the regulator knew the relative damages of pollution from those two places, she could institute a "trading ratio"—something like an exchange rate for allowances. A team of economists has estimated the system of exchange rates that would apply to the eight largest cities in the Upper Ohio River Basin, using an EPA water quality model to do so. Accounting for differences in the marginal damages from 1 kilogram of BOD load, Morgantown would need to purchase 4.28 kilograms of overflow abatement from Pittsburgh to obtain credit for 1 kilogram of its own abatement. If all trades took place at these estimated exchange rates, then the problem of hot spots from nonuniform mixing of sewer overflows could be solved.

The analogous tax is even more straightforward. Rather than applying the same tax to every source, the regulator could apply a tax more closely tied to the variation in marginal damages. For example, the regulator could charge 4.28 times more for a kilogram of overflow in Morgantown than in Pittsburgh. The point is that nonuniform mixing does not eliminate the case for market-based instruments per se. Rather, it points out the need to think carefully about the policy goal. If hot spots can arise, then a goal of reducing aggregate pollution may not make sense. If we redefine the goal, we can fashion a market-based policy to achieve it.

Monitoring and Enforcement

When we discussed cost-effectiveness, and even when we analyzed the incentives for technological change, we focused exclusively on abatement costs—that is, the costs to polluting firms (or individuals) from installing and operating pollution control equipment, and more generally from changing their production processes to reduce emissions (for example, by burning cleaner fuels). But in principle we care about minimizing *all* costs associated with environmental protection. After abatement, the prime source of costs is administrative—especially the costs of monitoring and enforcing compliance with environmental regulations.

For environmental regulations that target industries, these administrative costs are likely to be small relative to the costs of pollution abatement. For example, scrubbers to remove sulfur dioxide cost tens or hundreds of millions of dollars to install, and several million dollars per year to operate thereafter. The costs of emissions monitoring equipment are a hundred times smaller—a few hundred thousand dollars for installation, and on the order of fifty thousand dollars in annual operating costs. Emissions data is collected automatically by smokestack sensors and relayed electronically to computers at the Environmental Protection Agency, which also helps to keep administrative costs low.

Regulating the behavior of individuals is a different story—largely because of the sheer numbers involved. A year's worth of emissions and operation data from all of the fossil-fueled boilers at electric generating plants in the United States (there are 2500 of them) fits comfortably on a single spreadsheet. In contrast, there are 65 million households burning oil or natural gas to heat their homes. Or consider automobile emissions. With 140 million automobiles in the United States alone, the costs of monitoring and enforcing automobile emissions standards—quite apart from the costs of installing and maintaining emissions reduction equipment—make up a significant component of the overall cost of regulating automobile emissions.

One way to keep administrative costs down is to focus on technology and fuel inputs rather than the performance of individual sources of

Hot Spots and Pollution Trading: The Case of Nitrogen Oxides

A real-world hot spots problem is demonstrated by the cap-and-trade program for nitrogen oxides (NO_x) that has been functioning in the eastern United States since 1998.[9] NO_x combines with volatile organic compounds (VOCs), produced by both human and natural processes, to produce ground-level ozone—the main component of smog. While cars and trucks account for the majority of NO_x emissions, large stationary sources such as electric power plants are also important contributors to pollution. These stationary sources are the ones required to participate in the cap-and-trade program.

Importantly, the trading program operates as if each pound of NO_x causes the same damage, regardless of location; thus a power plant in North Carolina can trade on a pound-for-pound basis with a power plant in Maryland. The problem is that a pound of NO_x in North Carolina and a pound of NO_x in Maryland are altogether different: the former blows out to sea, while the latter contributes to urban smog in the heavily populated downwind areas of New Jersey and New York. Under a simple trading program, however, there is nothing to prevent the power plant in Maryland from emitting large amounts of NO_x and buying allowances from the North Carolina plant. Such a trade amounts to reducing air pollution over the Atlantic Ocean and increasing it over New York City—not a desirable exchange.

One possible solution (as suggested in the text) would be to define simple *trading ratios* between plants in different regions. An economist named Meredith Fowlie has simulated the gains from doing exactly that in the context of NO_x. She considered two particular policies. The first would impose a trading ratio of 1.5 to 1 between "high-damage" states (like Maryland) and "low-damage" states (like North Carolina). In other words, a power plant in Maryland would need 1.5 allowances for each pound of emissions, while a North Carolina plant would need only one. The second policy would impose a trading ratio of 5 to 1 between the same regions. Using a sophisticated statistical model, Fowlie estimated that the 1.5:1 trading ratio would induce a shift of 125 tons per day (about 6 percent of daily emissions) from high-damage to low-damage states. The 5:1 trading ratio, meanwhile, would move about 450 tons per day (22 percent). To put these figures in perspective, scientific studies indicate that shifting 11 tons per day from high-damage areas to low-damage areas, over a period of ten days, would save one human life.

pollution. This helps explain why command-and-control approaches tend to be prevalent in regulating home heating equipment, automobile emissions, and the like. Monitoring emissions from every household furnace would cost an astronomical amount—even before taking into account the administrative costs of levying an emissions tax, or the transactions costs involved in trying to institute a household-level market for pollution allowances. In such cases, technology standards (like those imposed on new oil and gas furnaces) or input standards (such as restrictions on how much sulfur and other contaminants can be present in fuel oil) make a great deal of sense.

A second issue related to monitoring and enforcement concerns the rates of compliance with environmental regulation. A sometimes overlooked advantage of market-based instruments is that they may *enhance* compliance. For example, transferable fishing quotas essentially give ownership of the resource to the people working the fishery—giving them incentives to help police their fellow fishers. If I pay for valuable rights to harvest from a fishery, I have a concrete financial incentive to make sure you don't catch more than you are allowed!

In the area of pollution control, command-and-control regulations have been undermined by ubiquitous delays, extensions, negotiated agreements with polluters, and the like. For example, the 1972 Federal Water Pollution Control Act required the EPA to announce regulations on effluent standards within a year. The EPA had still not completed the task by 1977—the date that sources were supposed to be in compliance with the regulation. That same year, Congress pushed back the date of compliance to 1984. Command-and-control regulation also raises the stakes for polluters, since installing new technologies can cost hundreds of millions of dollars. This gives them an incentive to sue regulatory agencies in an effort to stall or overturn environmental regulation—creating a major hurdle to the successful and timely implementation of those laws.

In contrast, the very nature of market-based instruments makes compliance easier, since firms have the option of paying a tax or buying allowances rather than installing new equipment or changing their practices. Remember that the total amount of pollution depends on the emissions tax or the pollution cap. If some firms opt to comply by paying for their pollution instead of reducing it, others will respond by doing much more than they would under command-and-control. Hence the flexibility built into market-based instruments can make pollution control easier to achieve and enforce, without leading to any more pollution.

Is Command-and-Control Ever Preferable?

Throughout our discussion of policy design, in chapter 8 and on into the current one, we have emphasized the arguments in favor of market-based instruments. While such approaches have many advantages—they can achieve an efficient level of pollution control, are cost-effective, and promote technological change—they are not panaceas. If you have been reading carefully, you will have noticed that in some settings command-and-control policies will be preferable.

For example, when hot spot problems are severe, command-and-control approaches may be the only feasible option. Regulating the disposal of toxic wastes requires restricting the amount of waste, and the means of disposal, at specific sites—not simply taxing firms for dumping waste, or imposing a nationwide limit on how much is created.

Occasionally, a certain control technology may be so effective and widely available, and production practices within an industry so similar across firms, that requiring the installation of that technology makes much more sense than regulating emissions. Double-hulled oil tankers are a good example. All shipping firms use essentially the same technology (large oceangoing tankers), meaning that there is likely to be little variation in abatement costs—and concomitantly little gain from allowing firms to trade control responsibilities among themselves. (This is compounded by the very high costs of cleaning up oil spills after the fact, and the difficulty of recovering those costs from the firm[s] responsible.)

Finally, as we saw in the case of administrative costs, the number of regulated entities may make market-based approaches impractical. Consider automobile emissions, for example. Catalytic converters (required on new cars sold in the United States since the mid-1970s) represent a classic technology standard; but they also are much more sensible than trying to regulate emissions from individual cars. This is partly because it is much cheaper to require automakers to install catalytic converters than to monitor emissions, and partly because the costs of abatement are unlikely to vary all that much among different car owners, limiting the gains from greater flexibility to comply with regulation. Even when automobile emissions are regulated, it is through performance standards based on periodic tests, rather than by taxes or cap-and-trade (which would require information on actual emissions).

Of course, technological advances can overturn these calculations of costs and benefits. Even ten years ago, it would have been astronomically expensive to monitor automobile emissions directly. But that may change in coming years. A pilot program in Oregon, for example, has set up remote monitoring sensors along roadways that can detect pollution from car exhaust pipes in real time, as the cars drive by. Paired with license plate recognition technologies, or sensors like those in automated toll booths that communicate with transmitters in the cars, such a real-time monitoring program could make an automobile emissions tax feasible. Perhaps in a later edition of this book we will be using car emissions as a standard example of where an emissions tax makes sense!

Conclusion

This chapter has focused on the choice between market-based instruments and command-and-control policies. We started by distinguishing between *goals* and *means* in environmental policy. This distinction is useful because it allows us to analyze the design of policy (should we require a certain abatement technology or create an allowance market?) independently from the policy target (how much pollution should we allow?).

Regardless of how the policy goal is set, market-based instruments have two substantial advantages over technology and performance standards. First, they are cost-effective: A tax or cap-and-trade program can (at least in theory) achieve a given goal at least total cost. They do this by ensuring that all firms end up with the same costs on the margin—since all firms face the same incentives, in the form of the tax or the allowance price. In contrast, a uniform performance standard (and to an even greater degree a technology standard) is a one-size-fits-all approach that ignores the different opportunities available to different firms. While firm-specific performance standards would be cost-effective in theory, they are likely to be infeasible in practice, due to the enormous informational requirements they impose on the regulator.

The second major advantage of market-based instruments is that they promote the adoption and innovation of new technologies. If firms pay for every ton of pollution they emit, for example, they have a strong incentive to look for new and better ways to reduce pollution.

There are of course a number of other arguments for (and against) market-based instruments. These relate to the possibility of hot spots, heterogeneity among regulated firms, and the costs of compliance and enforcement. While market-based instruments are not the answer to all environmental problems, they do represent a very powerful and widely applicable component of the environmental policy toolkit. In the next chapter, we shall see how they have been implemented in the real world.

10

Market-Based Instruments
in Practice

We have now seen how environmental degradation and excessive resource extraction are often due to absent or incomplete markets and how the best solutions to such market failures may involve market principles themselves. Now we look at some examples of how market-based policies have been used in the real world.

We begin by describing three cases in detail: the market for sulfur dioxide (SO_2) emissions from power plants in the United States, the tradable individual fishing quota (IFQ) system for New Zealand's fisheries, and municipal drought pricing of water resources in the United States. In each of these cases, we describe the background to the regulations, discuss their performance (both in meeting policy goals and in achieving cost savings relative to conventional regulatory approaches), assess their distributional effects, and touch briefly on questions of compliance and enforcement. We then describe a number of applications of market-based policies to environmental and natural resource management—ranging from water quality trading, to "pay as you throw" programs to manage household trash disposal, to "banks" for wetlands and endangered species.

The purpose of the chapter is not to catalog every application of market-based approaches to environmental management and document successes and failures.[1] Instead, we demonstrate specific cases in which market principles have been applied by governments to correct market failures. In some cases, these attempts have had great success, in others, less so. We hope to leave you with the capacity to think broadly and creatively about the ways in which market principles can be used to promote environmental protection by aligning the incentives of firms and consumers with the interests of society.

The U.S. Sulfur Dioxide Market

One of the most successful market-based environmental policies has been the U.S. sulfur dioxide allowance trading program set up by the 1990 Clean Air Act. This program, more than any other, has shown that market principles can improve environmental regulation, and has served as a model for other environmental markets.

To understand how the program works, it helps to start by describing the run-up to the eventual legislation. Throughout the 1980s, Congress debated reauthorization of the Clean Air Act, which had last been updated in 1977. One impetus was the growing concern over acid rain, largely caused by sulfur dioxide from power plants in the Midwest. Many of these power plants had been built in the 1950s or 1960s. As a result, they had been grandfathered into the laws passed the 1970s, which only covered new sources. Indeed, the continuing role of these large power plants in emitting pollution exposed a fundamental flaw in the earlier approach. The logic behind the decision to focus on new sources was simple enough: Installing pollution control equipment at a new source is much less expensive than retrofitting existing plants. Over time—so the theory went—the new standards would cover a greater and greater fraction of electricity generation, as existing plants were retired and replaced with new ones. But by making new power plants much more expensive to build, Congress had unwittingly made it much more attractive to keep old power plants in service. As a result, plants that were originally scheduled to last thirty years were still going strong, with no retirement in sight.[2]

If Congress were to start regulating these older plants, how should it do so? Earlier clean air legislation (in 1970 and 1977) had adopted command-and-control regulations, first setting a uniform emissions standard, and then imposing a technology standard requiring power plants to install scrubbers. By the late 1980s, however, the idea of emissions trading was moving from academic journals to the policy world. Officials at the Environmental Protection Agency (EPA) had begun to introduce market principles, letting polluters offset increased pollution at one facility with reductions in pollution elsewhere. The new ideas were even gaining ground with some environmental advocates—in particular, the Environmental Defense Fund (EDF, now simply Environmental Defense), whose director, Fred Krupp, described market-based policies as a coming "third wave" of environmental policy.

The 1990 legislation resulted from an unlikely alliance. On one side was the administration of President George H. W. Bush, who had campaigned in part on a pledge to be "the environmental president"—going so far as to film a campaign spot near heavily polluted Boston Harbor to undercut his rival, Governor Michael Dukakis of Massachusetts. At the same time, however, Bush needed to promote his bona fides as a business-friendly Republican. On the other side of the alliance was EDF. The two sides essentially struck a deal. EDF would endorse (indeed, would help to write) legislation proposed by the Bush administration that enshrined a market-based approach to SO_2 control—helping to give the Bush initiative credibility among moderate environmentalists. In exchange, the administration proposal would set of goal of reducing SO_2 emissions by 10 million tons per year from 1980 levels by the year 2000, rather than a weaker target of 8 million tons.

Congress passed the resulting legislation (with some modifications) as Title IV of the 1990 Clean Air Act Amendments. It contained two particularly notable provisions for sulfur dioxide regulation. For the first time, Congress exerted federal authority over emissions from plants built before 1971. And in bringing these existing power plants under federal regulation, Congress adopted a novel tack: a system of tradable allowances (or cap-and-trade policy). In the first phase of the program, which lasted from 1995 through 1999, 263 generating units at 110 power plants were required to participate in the allowance market. Although these made up less than one-fifth of the total number of fossil-fired generating units, they were by far the largest and dirtiest; environmental advocates even referred to them as "The Big Dirties." Total pollution in this phase was capped at roughly 6.3 million tons of SO_2 per year (an annual reduction of around 3.5 million tons). This cap was divided up into the same number of allowances, each corresponding to one ton of pollution, which were allocated for free to the regulated power plants. In the second phase of the program, starting in the year 2000, virtually all power plants above a certain size were brought into the allowance market, with an overall annual cap at 9 million tons.[3]

Power plants were allowed to bank their allowances for later use. An allowance handed out in 1996, for example, could be used in that year, or saved for a later date. (The reverse was not true: allowances could not be borrowed from future years.) As we saw in chapter 8, this feature provides additional flexibility for regulated firms to reduce their pollution over time.

Performance

The SO_2 allowance market has been widely hailed as a tremendous success. Market activity has been substantial, with the annual volume of trades equaling or exceeding the total number of allowances allocated each year. Several brokerage companies compete to track and arrange bilateral trades. Forward markets, loans, swaps, and other financial derivatives have also sprung up. Meanwhile, the market has not only succeeded in meeting the cap, but achieved *more* abatement than required in the first phase of the program. These early reductions, made possible by the banking provision, were encouraged by the effective tightening of the cap starting in the year 2000. Looking ahead, electric utilities foresaw that allowance prices would rise (and abatement would become more costly) as allowances became more scarce in later years. In response, utilities as a whole abated more than they were required to—roughly 2 million more tons per year than required, in fact, amounting to about one third of the total allocation. After the start of Phase II, they started to draw down the resulting allowance bank. From an economic point of view, this early abatement is a good thing, since more benefits from lower pollution were enjoyed earlier than would have been the case under command-and-control regulation with a similar target.

To assess the economic performance of the Title IV trading program, we can start by comparing the estimated costs and benefits. The major categories of benefits from sulfur dioxide abatement include lower incidence of sickness and mortality due to urban air pollution; reduction in the acidification of aquatic ecosystems; and impairment of visibility in recreational and residential areas. Economic analyses have found that the lion's share of the benefits turn out to come from health effects, rather than the ecological impacts of acid rain. Indeed, one authoritative analysis found that on a per capita basis for the northeastern United States, reduced sickness and mortality accounted for over 85 percent of the benefits.[4]

Moreover, total benefits outweighed estimated costs by roughly an order of magnitude. At a national level, the same study estimated that health benefits to the United States as a whole were $3,300 per ton of sulfur dioxide reduced—compared with costs of around $270 per ton. An interesting irony arises here. The impetus for reducing sulfur dioxide pollution was concern about acid rain and the attendant damage to ecosystems—not about human health impacts. At the time the legislation was written and passed, the contribution of sulfur dioxide emissions to urban air pollution was not widely understood. Indeed, the trading program itself was dubbed the "Acid Rain Program." If the benefits had been limited to reducing the ecological impacts of acid rain, however, the costs would have been comparable in magnitude to the benefits, with little net gain to society. This is a happy case of a positive unintended consequence. From an efficiency perspective, the allowance trading program did the right thing for the wrong reason.

Another way of assessing the program's performance is suggested by our discussion in chapter 9. Taking the goal of the policy as given, how well did the cap-and-trade policy that was actually used perform, relative to alternative policies that might have been employed instead? Table 10.1 provides some answers to this question, at least for Phase I of the sulfur dioxide trading program. Three counterfactual scenarios (what could have happened but didn't) are given, along with the baseline scenario corresponding to the actual cap-and-trade program. As the table shows, the cap-and-trade program was considerably more costly than the theoretical minimum cost of achieving the same emissions reduction. In other words, if electric utility managers had had perfect foresight about the prices of low-sulfur coal and SO_2 allowances, and had made the cost-minimizing decisions based on that information, total abatement costs would have been less than half of what they actually were. This is a useful reminder that policies in the real world never work as perfectly as they do in theory.

However, it is more informative to compare the outcomes of the actual policy with estimates of what the outcomes would have been from other feasible policies—not just the theoretical best-case scenario. The table also shows that the cap-and-trade program was significantly less costly than a uniform emissions rate standard on the same power plants. The estimated savings of just over $150 million represent about one-fifth of the costs of the actual trading program. These cost savings, in turn, pale next to the cost savings relative to a technology standard requiring utilities to install scrubbers. Such a policy (along the lines of the 1977 Amendments) was actively considered in the debates leading up to the 1990 legislation. If it had been chosen, the costs of achieving the same amount of abatement would have been over three times as high as they were: nearly $2.6 billion per year, compared with a baseline annual cost of $747 million. These large cost savings are due in large part to wide variation in abatement costs among electric utilities. For example, much of the steep reduction in sulfur dioxide emis-

TABLE 10.1
Estimated Costs of Various Alternative Policies to Achieve Same Emissions
Reduction in Phase I of the 1990 Clean Air Act Amendments

Scenario	Estimated annual cost (millions)	Cost difference from baseline (millions)	Cost increase
Theoretical least-cost outcome	$315	–$432	–57%
Baseline cap-and-trade program (actual policy)	$747	—	—
Uniform emissions rate standards	$900	$153	20%
Technology standard	$2,555	$1,808	242%

Estimates taken from Nathaniel O. Keohane, "Cost Savings from Allowance Trading in the 1990 Clean Air Act," in Charles E. Kolstad and Jody Freeman, eds., Moving to Markets in Environmental Regulation: Lessons from Twenty Years of Experience (New York: Oxford University Press, 2007).

sions under the program has been due to greater use of low-sulfur coal from the Powder River basin in eastern Wyoming. Transportation (mostly by railroad) accounts for most of the cost of extracting and delivering coal to midwestern power plants. As a result, a power plant's geographical location is a key determinant in its abatement cost.

We can also frame cost-effectiveness another way—in terms of the pollution savings. Recall the role played by the environmental group EDF in supporting the allowance market. To a group like EDF, the cost savings from market-based policies offered an opportunity to secure greater environmental protection. The standard case made by economists (indeed, the case we made in chapter 9) emphasizes cost-effectiveness: Market-based policies can achieve a given policy goal at less cost than command-and-control. But such policies can be portrayed equally well another way: For a given total cost, a cap-and-trade policy allows for more abatement. In the case of sulfur dioxide, the use of a cap-and-trade program translated into roughly 10 percent more abatement being done than could have been done for the same total cost with a uniform emissions rate standard.

Finally, we can consider the effects of the cap-and-trade program on technological change. While it is still too early for a full assessment of how the program has affected the development of new abatement technologies, some early prognoses have been offered. In line with what we would expect, based on chapter 9, allowance trading seems to have boosted the incentive for electric utilities to adopt lower-cost technologies. Among electric utilities regulated by conventional emissions rate standards, the cost of scrubbing appears to have had little impact on the decision to install a scrubber. In contrast, cost was an important consideration at power plants included in the allowance-trading program. Evidence from patent data, meanwhile, suggests that the SO_2 trading program spurred firms who design and build scrubbers to focus more on raising removal efficiencies than they had under previous command-and-control regulations (which did not reward increases in abatement beyond that required by the uniform standard). Both findings are consistent with theoretical predictions that market-based policies promote greater technological innovation.[5]

Distributional Implications

In the case of a cap-and-trade program like that for sulfur dioxide, we can assess the distributional impacts along two very different dimensions: how were the allowances allocated, and where did the pollution end up?

At a broad level, the entire 6-million-plus allowances (during Phase I), as well as the 9-million-ton cap in Phase II, were given away for free to existing power plants. An alternative allocation mechanism would have been to auction the allowances.[6] As we discussed in chapter 8, an auction would have raised government revenue that could have been used to offset distortionary income and sales taxes. Instead, that revenue was essentially handed to the electric power plants that participated in the market. In effect, the 1990 CAAA created a new scarce resource, and gave away the rents. A quick back-of-the-envelope calculation can give you a sense of the total value of the scarcity rents involved. The average allowance price during Phase I was $135; multiplying by 6.3 million allowances implies a total annual value of $850 million. We saw above that the estimated annual cost of abatement was about $750 million—implying rents on the order of $100 million a year.

What about the consequences of allowance trading for the distribution of pollution? As we noted in chapter 9, a potential concern with market-based instruments arises when pollutants are not uniformly mixed. As it turns out, SO_2 is not uniformly mixed. Emissions from power plants in Ohio and Indiana travel downwind to the urban centers of the Northeast, and contribute to acid rain in the Adirondacks. Emissions from power plants in Maryland or Delaware, on the other hand, are much more likely to blow out to sea. In theory, the SO_2 trading program could have made matters worse—even as it lowered total pollution—if it had led to the reallocation of pollution from Delaware to Ohio. Indeed, this issue was voiced by critics of the program.

Although the jury is still out on whether these concerns have materialized, the evidence so far suggests that they have not. In fact, just the opposite seems to have occurred. By far the largest reductions in emissions have taken place in Ohio, Indiana, and other midwestern states—precisely those states whose emissions were of greatest potential concern.[7] The reasons for this are varied, but three stand out. First, the required abatement implied by allowance allocations were largest in these regions, since they were also the biggest emitters before the program started. In other words, even if electric utilities had simply used up the allowances given to them, without trading among each other, abatement would have been greatest in these midwestern states. Second, abatement was cheaper for many midwestern plants than for other power plants. Relative to power plants in West Virginia, Georgia, and farther east, midwestern plants have easy access to low-sulfur Wyoming coal. Finally, state-level regulations may have played a role. Even after the introduction of the national trading program, some states still imposed limits on pollution from individual power plants, in order to protect local air quality. And states like Ohio and Indiana sought to protect their in-state high-sulfur coal industries by encouraging power plants that participated in the allowance market to install scrubbers. As this discussion makes clear, there was nothing in the cap-and-trade program itself that prevented hot spots from being a problem. Rather, it was a combination of factors, including dumb luck, that made the difference.

Compliance and Enforcement

Compliance and enforcement have not posed major hurdles for the SO_2 trading program. We have noted already (in chapter 9) that the costs of monitoring, while real, have been roughly two orders of magnitude less than the costs of abatement. Moreover, these monitoring costs would have been incurred by any approach based on emissions—for example, a uniform performance standard.

Observers typically note that compliance with the program has been "100 percent"—meaning that participating utilities have indeed retired as many allowances as needed to cover their emissions. Since the program allows for a trueing-up period early each year (firms have until March 1 to ensure that their allowance holdings cover their previous year's emissions), while utilities are fined two thousand dollars per ton for noncompliance (far above prevailing allowance prices), this high rate of compliance is hardly surprising. Nonetheless, the smooth experience with the allowance trading system may have broader implications for market-based policies. As we mentioned in chapter 9, by reducing the overall cost of complying with regulations, market-based instruments may help reduce the incentives for polluters to challenge regulations in court. In sharp contrast to earlier command-and-control regulations—including the restrictions imposed by the 1970 and 1977 Clean Air Act Amendments—there have been no major lawsuits challenging the legality of the allowance trading program or delaying its deadlines.[8]

Individual Tradable Quotas for Fishing in New Zealand

In 1986, New Zealand set up what has become the world's largest market for tradable individual fishing quotas (IFQs). At the time, concerns about overfishing created a sense of crisis—especially for an island nation heavily dependent on its natural resources. A market-based approach to fisheries management also appealed to the government at the time, which pursued a larger agenda of market reform, including privatizing industries in some sectors and reducing subsidies in others.

Each year, the government sets a total allowable catch (TAC) for each species-region, based upon a biologically determined maximum sustainable yield (MSY).[9] Quotas are freely allocated based on a fisher's average catch in a set of preceding years. Fishers can buy and sell IFQs, which represent the right to fish a percentage of the TAC in perpetuity.[10] With few exceptions, these quotas cannot be traded across regions, species, or years. When the system began, it covered twenty-six different species. By the mid-1990s, IFQ markets covered 85 percent of the commercial catch within the exclusive economic zone extending two hundred miles out to sea from New Zealand. By 2004, seventy different species were included in the program. Moreover, coastal waters are spatially divided into quota management regions, with markets in each region for each relevant species—generating more than 275 separate markets.

Performance

The first question about the performance of an IFQ market is: Did it help to reduce overfishing and restore the stocks? From its very beginning, the program mandated a sharp reduction in the catch. In 1986, the TACs were only a quarter to three-quarters of what they had been before the program started. (Where a particular fishery fell within that range depended on biological status and management goals.)[11] Available evidence suggests that few if any species populations are worse off as a result of the IFQ system (especially in comparison to the alternative of continued open access), and some show significant positive signs of recovery. One analysis of the rebound of regulated fish stocks in New Zealand examined 149 of the 179 individual stocks (defined by species and region) that were governed by the system in 1993.[12] Only thirteen of these stocks were estimated to be smaller than the stock that would generate the maximum sustainable yield. Another thirteen were estimated to be at or above this level, while the status of the remaining seventy-five was not determined. TACs for those fish stocks deemed to be below the MSY-supporting level have been reduced over time. This performance stands in contrast to the well-publicized crashes of fish stocks in other parts of the world over the same period of time, foremost among them the sharp decline in the North Atlantic cod fishery.

As with the SO_2 cap-and-trade program, we can also assess the performance of the New Zealand IFQ program relative to alternative command-and-control policies, such as nontradeable permits. No study has estimated the cost savings directly. But an analysis of market activity and IFQ prices can tell us a great deal about how well the market is performing.[13] A recent study of the New Zealand IFQ market found that 70 percent of quota owners had taken part in a market transaction by 2000. Quota sales were highest early in the program, as the initial allocation was redistributed among fishers. This is consistent with increased efficiency: Presumably, more profitable producers bought their way into the program, while less profitable producers sold their quota and exited.

The variation of prices around their annual mean can also tell us something about how smoothly the market is operating. In a well-functioning market, the price of an IFQ would reflect the present value of the stream of expected future net revenue from future harvests. Economic theory would predict a single price within a species-region, just as in any well-functioning competitive market. In reality, transactions costs and other imperfections can lead to price variation even for an identical good. (Think, for example, of gasoline prices—which often vary widely within a neighborhood despite the fact that the underlying product is essentially identical.) In the New Zealand IFQ market, quota prices varied by as much as 5 percent around their mean value, comparable to price dispersion in other real-world markets. Moreover, price dispersion has been trending downward since the market's inception in 1986—suggesting that market frictions are diminishing.

Trends in quota prices give a measure of the long-run performance of the program. A prime goal of fisheries management, after all, is to raise the economic value of the fishery. Remember the basic intuition that IFQs work by establishing property rights to the fishery. Just as house values rise with neighborhood revitalization, the value of a fishing quota should rise over time as the health of the fishery improves. This is exactly what appears to be happening in the New Zealand market. Quota prices, adjusted for inflation and controlling for a range of outside factors, have risen at a rate of 5 to 10 percent over the course of the program, with higher rates observed in markets that saw greater initial reductions in the allowable catch and correspondingly higher consolidation of quota ownership.

Distributional Implications

One of the strongest objections to IFQ markets is that they encourage consolidation of the fishery—in other words, a reduction in the number of fishers operating in the market. This occurs partly because of the overall decline in the allowable catch that often accompanies a new market, just like under any fisheries regulation. But consolidation tends to be greater under an IFQ system, which allows less profitable fishers to exit the industry, while more profitable fishers enter or expand to capture a greater share of the market. From an efficiency perspective, the exit of high-cost fishers and their replacement by low-cost fishers is exactly the purpose of the market. After all, falling costs boost the economic value of the resource, leading to higher profits for the fishers who remain in the industry, lower prices for consumers, or both.

The implications of consolidation for equity, however, are mixed. Critics argue that IFQs give an unfair advantage to large firms that can take advantage of scale economies to reduce their fishing costs below those of small-scale fishers.[14] They also lament the decline of a way of life that in many cases has been handed down over several generations. On the other hand, it is not clear that the distributional implications are negative. After all, any fishers who exit the fishery do so by choice, and receive profit from selling their quota as a result. In contrast, restricting harvests without allowing trading may deny fishers the choice of whether to remain or exit, and those who exit do so without compensation. Indeed, any adverse distributional impacts of IFQ systems must be compared against the consequences of some other form of restriction on harvests (which will also harm fishing communities) or of continued open access (which is likely to bring on the collapse of the fishery).

IFQ markets can be designed to accommodate some of these concerns. For example, the initial allocation of quota provides a way to address distributional issues. Recall from our discussion of tradeable pollution allowances in chapter 9 that the *equilibrium* allocation of quota in a market is essentially independent of the initial allocation (as long as transactions costs are low). If equity is a concern, fisheries managers can hand out a disproportionate share of quota to small-scale fishers, or those who have had a long history fishing the resource. If consolidation *per se* is a concern, it can be limited by policies that impose maximum quota holdings or the like. However, unlike changes to the initial allocation, such constraints on the final allocation of quota come at a substantial cost in terms of efficiency.

How does all this play out in the case of New Zealand? Some consolidation has certainly resulted. Between 1986 and 2003, the number of quota owners in the IFQ system declined by 37 percent, primarily in species-regions most severely overcapitalized before the introduction of markets.[15] Many of those who have exited the markets are small-scale fishers, but many of these remain, as well.

Another distributional issue in New Zealand's IFQ system concerned its impact on the traditional fishing communities of native New Zealanders, the Maori. The initial design of the IFQ market excluded the Maori altogether. In response, the Maori successfully sued the government in the late 1980s, arguing that they had been disenfranchised. As a result of this litigation, the government allocated shares of the TAC to Maori communities in 1992 (10 percent of all existing IFQs, plus 20 percent of the TAC for any new fish stock regulated by the program), and provided funding for the purchase by Maori of one-half of New Zealand's largest fish company.[16]

Compliance and Enforcement

Traditional fishing regulations are famously difficult to monitor and enforce. For example, limiting fishing seasons or particular fishing areas results in steep increases in the concentration of effort within allowable times and places. Limits on the types of allowable fishing technologies create strong incentives for the development of alternative fishing gear that is equally, and in some cases even more, productive. Gear restrictions require on-boat inspections, and the enforcement of fishing seasons and area restrictions requires that regulators monitor fishing activity at sea.

Like traditional fisheries regulation, market-based regulation requires monitoring and enforcement to ensure compliance. Within New Zealand's IFQ program, quota holders pay levies per metric ton of quota share to support the cost of managing and administering the program, including enforcement costs.[17] Enforcement under an IFQ program is mainly a matter of auditing paperwork and the exchange of fish between boats and fish purchasers where fish are landed. Penalties for violating the IFQ regulations in New Zealand include forfeiture of fishing quotas, seizure of property, and exclusion from the fishing industry.

In comparison to what may occur under traditional regulations, fishers' attempts to maximize the value of their catch, given their quota, under a market-based system can lead to practices like high grading and price-dumping. High grading involves the discard of lower-valued members of the quota species (for example, smaller fish), so as to fill the quota with the most profitable fish. High-grading of snapper was a problem in the early years of New Zealand's IFQ program, but vigorous enforcement has ended the practice.[18] Price-dumping occurs when fishers discard their catch when fish prices fall, to leave room to fill their quota on more profitable days when prices are higher. It is difficult to determine how frequent such practices are. One analyst surveyed New Zealand fishers in 1987, at the start of the program, and found that 40 percent thought that enforcement of the IFQ system was an important problem, while 66 percent were concerned about high-grading. By 1995, however, the comparable figures were 21 and 25 percent, respectively.[19]

A related concern involves the effects of fisheries management on other marine life—a problem known as by-catch. Fishers concentrating on a particular valuable species often end up catching other unwanted fish or marine animals in their nets. (The most famous example is the dolphins caught in seine nets formerly used to catch tuna in the eastern Pacific.) There is reason to believe that the comprehensive nature of the New Zealand IFQ system may indirectly alleviate the by-catch problem. The quota markets cover so many species that fishers have taken significant measures to reduce by-catch, in order to avoid the need to purchase additional quota, or to pay fees to regulators for species caught outside of a fisher's quota portfolio.

Nonetheless, by-catch is a significant issue in some New Zealand fisheries. The important question is how by-catch under a market-based system compares to that which we would observe under traditional fishing regulation. As yet, the data from New Zealand are insufficient to make this comparison.

Municipal Water Pricing

Cities in arid U.S. states such as Texas and California have struggled to manage water scarcity in the face of growing populations, high demand for household water use (for swimming pools and landscaping, among other things), and the increasing cost of acquiring and developing new water supplies. During droughts, cities typically implement voluntary and/or mandatory limits on residential water consumption. For example, cities may restrict certain activities such as watering lawns, or require homeowners to install water-saving devices such as low-flow shower heads. As you have probably realized by now, these kinds of blanket limitations on water consumption are not how an economist would approach the problem of managing scarce water resources. What do market-based solutions look like in this context?

A key economic insight from our discussion of natural resources in chapters 6 and 7 is that the price of a resource should reflect its scarcity. Setting higher water prices during droughts (when scarcity is greater) could achieve the same reductions in water use as the kinds of policies mentioned above, at lower total cost. Moreover, such prices promote the efficient allocation of water among competing demands. Higher prices would ensure that scarce water flowed to the highest-valued uses. Relative to one-size-fits-all restrictions, consumers with high willingness to pay for water would end up consuming more, while consumers with low willingness to pay would cut back on their use. We have emphasized the critical role played by "cost heterogeneity" in pollution control or fisheries management. In the context of water pricing, "benefit heterogeneity" plays an analogous role. The greater the differences between water consumers, and the more varied their marginal willingness to pay for the resource, the greater the gains will be from allocating water by prices or markets.

Performance

As a baseline for assessing the performance of drought pricing, let's start by considering the effectiveness of command-and-control alternatives. Here, the evidence is mixed. In general, requiring customers to install specific water-conserving technologies does reduce consumption—but typically by much less than the "manufacturing specifications" for the conservation technology would predict.[20] (One possible reason is that consumers change their behavior after installing the new technologies—for example, starting to take longer showers after installing a low-flow showerhead.) A comprehensive study of various conservation programs in California found that public information campaigns, retrofit subsidies, water rationing, and water use restrictions all contributed to reductions in residential water use, with the more stringent, mandatory policies having stronger effects than voluntary policies and education programs.[21]

Price increases have been used infrequently as a drought management tool. During an extended drought in California from 1987 to 1992, a handful of municipal water utilities implemented price increases to reduce water demand, achieving aggregate demand reductions of 20 to 33 percent with very substantial price increases.[22] In principle, it would be straightforward to design a drought pricing system that would result in the same aggregate water savings as an existing nonprice approach. The key piece of information needed in each case would be the price elasticity of water demand— a measure of how sensitive consumers are to changes in the price of water.[23] Estimates of residential water price elasticities usually range from −0.3 to −0.6, meaning that a 10 percent price increase can be expected to reduce demand by between 3 and 6 percent.

In the absence of extensive empirical experience, analysts have used data on actual water use to run simulations of what would happen under hypothetical (but plausible) market-based policies. A recent study of thirteen California cities found that under a wide range of assumptions, a modest water tax (a price increase) would be more cost-effective than a technology standard (a mandatory low-flow appliance regulation).[24] Another study of eleven urban areas in the United States and Canada compared residential outdoor watering restrictions to drought pricing.[25] For the same level of aggregate demand reduction as that implied by a regulation allowing households to use water outdoors (for watering lawns and washing cars) only two days per week, the establishment of drought pricing in each city would result in welfare gains of approximately eighty-one dollars per household per summer drought. This represents about one-quarter of the average household's total annual water bill in this study.

In the long run, higher prices for water will lead to land-use patterns, investments, and consumption decisions that take account of water scarcity.

For example, we would expect households to plant fewer green lawns and install front-loading washing machines (which use much less water than top-loaders) relatively more often in cities where water prices are high. As in the case of market-based policies to reduce air and water pollution, prices provide a strong incentive for technological change that lowers the marginal cost of water conservation.

Distributional Implications

The main distributional concern with a price-based approach to urban water management arises from one of the central features of a market. To an economist, one of the virtues of markets is that they allocate resources according to who is willing to pay the most for them. But willingness to pay strikes some people as an unfair criterion for allocation, since it is strongly influenced by ability to pay. What you are willing to pay for something depends in part on how much money you have to spend. (Recall our discussion of willingness to pay versus willingness to accept in chapter 3.) This sense of unfairness may be especially acute when we are dealing with resources that satisfy basic needs—like water for drinking and bathing. These distributional impacts are illustrated by the results of the study of U.S. and Canadian cities mentioned above. Not surprisingly, raising prices during a drought, rather than restricting consumption, would result in more water being consumed by wealthier households with large lots, and less by poorer households. In welfare terms, as well, poorer households would be hit harder: The reduction in consumer surplus as a percentage of income is larger for low-income households.

Concerns about distributional equity could be addressed by combining drought pricing with income transfers. Higher water prices would yield substantial profits for the utilities—profits that would have to be returned to consumers in some form (since water utilities tend to be subject to strict price regulation and oversight). In the case of residential water use, higher prices during droughts could be offset by rebates (unrelated to water use) to low-income households, which could appear automatically on residential water bills. (Note the parallel with the use of IFQ allocations to meet distributional equity goals in the case of fisheries management.)

Compliance, Monitoring and Enforcement

Applying market principles to urban water management raises fewer concerns for monitoring and enforcement than do existing command-and-control approaches. Under outdoor watering restrictions during a drought, utilities rely on neighbors to report illegal watering—a notoriously ineffective system. Additionally, requirements for the installation of indoor water-saving fixtures cannot be monitored and enforced without utility representatives entering private homes. In contrast, drought pricing involves simply changing the price of water. Since almost all households pay according to how much water they use, and their water consumption is already metered, there's no need for additional monitoring and enforcement. Indeed, cheating in the drought pricing scenario would require households to figure out how to consume water off the meter. Of course, where metering is not prevalent, a drought pricing policy would require meter installation for all users, but so would any attempt to charge people according to the amount of water they consume.

Water Quality Trading

A handful of water quality trading programs involving trades among point source polluters have been established around the world. The program that most closely resembles the air quality trading programs we have discussed thus far is the salinity trading program in Australia's Hunter River. The Hunter River trading program, established as a pilot program in 1995, issues tradable permits to coal mines and power plants to discharge saline water into the river during periods of high flow, when dilution is greatest. Like tradable fishing quotas, salinity permits entitle polluters to emit a share of the total allowable discharge (rather than a specific quantity). Since flow conditions can change rapidly, trading is accomplished online, in real time, through a central Web site. The alternative for participating sources would have been the construction and maintenance of larger saline water reservoirs, at significant expense.

Dozens of water quality trading programs have been established in the United States over the past two decades, yet few active markets are in place.[26] Studies imply substantial potential cost-savings from water quality trading, especially between regulated point sources (who have climbed up their respective marginal cost curves due to three decades of regulation) and unregulated nonpoint sources. For example, in Idaho's Lower Boise River, the estimated marginal costs of reducing phosphorous loading from municipal wastewater treatment plants range from five dollars to over two hundred dollars per pound. In contrast, estimated marginal costs for equivalent reductions in the agricultural sector (through the implementation of improved land management practices) range from five to fifty dollars per pound.[27]

One of the more successful U.S. programs is a nutrient-trading program established in the early 1990s in North Carolina's Tar-Pamlico River Basin. Point sources and municipalities purchase agricultural nutrient reduction credits (as well as wetland and riparian buffer restorations) through an intermediary, the Tar-Pamlico Basin Association. The offset price set by the Association for these transactions in 1999 was twenty-nine dollars per kilogram of nitrogen or phosphorous. In comparison, estimated marginal costs for reductions by individual point sources ranged from fifty-five to sixty-five dollars per kilogram. By 2002, total phosphorous and total nitrogen concentrations in the basin had been reduced significantly.

U.S. water quality trading programs have been hindered by "thin" markets (many watersheds contain a fairly small number of point sources, in particular) and by the difficulties of verifying reductions at nonpoint sources, among many other factors. Water quality is a more difficult problem to address through markets than air quality, due primarily to the fact that most water pollutants are highly nonuniformly mixed, a term we introduced in chapter 9. Regulators must establish complex matrices that describe the effects of effluents by each potential polluter on pollution concentrations at various points within a water body. For example, say that we are trying to reduce the "dead zone" in the Gulf of Mexico, and we have determined that the two main causes are farming upstream of the Mississippi Delta and municipal sewage treatment plant discharge from cities in the delta. The effects on water quality in the Gulf of these upstream and downstream effluents, of nitrogen and phosphorous in different forms and quantities, will differ by source, and even by season. We cannot simply set a cap on the two pollutants in the delta region and allow polluters to trade, without first thinking about the ratios at which each pair of sources should be allowed to trade. For this reason and others, markets may hold less promise as a fix for some water quality problems than they do for carbon dioxide emissions, or for fisheries management.

Waste Management: "Pay As You Throw"

Market-based approaches have also been used to manage solid waste. Some waste products have high recycling value. We rarely observe piles of copper piping set at the curb by households on trash collection days. Similarly, Alcoa (the aluminum giant) sponsors local drop-off centers for aluminum recycling—not as a charitable effort, but because it is much less costly to produce aluminum from scrap metal than from virgin ore. The bulk of municipal solid waste, however, eventually ends up as trash, with legal and illegal disposal costs (like pollution externalities, collection costs, and landfill space) borne by the community as a whole.

One estimate suggests that the marginal cost of garbage collection and disposal by the public sector for an American household is $1.03 per bag. But until recently, the private marginal disposal cost for these households was zero.[28] Prompted by rising landfill costs and incineration fees (in part due to widespread community opposition to the siting of new waste disposal facilities "in their backyards"), many U.S. communities in the 1980s and 1990s began experimenting with volume-based waste disposal charges, also known as "pay-as-you-throw" systems. These market-based approaches, by internalizing some marginal waste disposal costs, create incentives to minimize waste volume through recycling, composting, and reducing demand for products with excessive packaging. The programs can take many forms, including: requirements for the purchase of official garbage bags, or of stickers to attach to bags, not to exceed a specified volume; periodic disposal charges for the number and size of official city trash cans collected at the curb; and charges based on the measured weight of trash. In 2003, more than four thousand U.S. communities had implemented some form of pay-as-you-throw disposal.[29]

Evidence is mixed regarding the effectiveness of these programs. For example, a study of a program in Charlottesville, Virginia, found that charging eighty cents per garbage bag resulted in a 37 percent decrease in the number of bags.[30] This effect, however, was offset by two factors that have proven to be endemic to such programs. First, the reduction in the weight of trash (a better indicator of disposal cost than volume, since garbage trucks compact trash bags anyway) was much smaller: only 14 percent. This phenomenon is known as the "Seattle Stomp," after the supposed exertions of Seattle residents who responded to volume-based charges by compacting their trash themselves. Second, charging for trash creates an incentive for illegal dumping. In the Charlottesville case, the true reduction in weight was only 10 percent once illegal disposal was taken into account. A broader study of twenty U.S. metropolitan statistical areas also found that the effects of unit disposal pricing are unclear, and almost certainly much smaller than the effects of simply providing curbside recycling pickup.[31]

Habitat and Land Management

Some of the most novel applications of market-based policies have concerned land-use management and habitat preservation. We discuss three prominent examples along this relatively new frontier: tradable development rights, wetlands mitigation banking, and habitat trading for endangered species preservation. At the end of this section, we discuss some common concerns that arise in this realm of market-based policy.

Tradable Development Rights

In developing countries, conflicts between conservation and development goals can be particularly intense. For households mired in poverty, the short-term cost of land-use restrictions can be very high. Conversion of forests and other ecologically valuable lands to agriculture is pervasive. Attempting to balance these concerns, some countries have implemented a variety of market-based policies to reweight relative returns to land uses, making the socially desirable land uses more competitive. For example, since 1965, land in southern Brazil has been subject to a requirement that each parcel of private property remain in native or regenerated forest, with mixed results. Recent changes exempt some landowners from this requirement, allowing them instead to offset the loss of forest to development on one parcel by preserving another parcel elsewhere. This policy, known as "tradable development rights," (TDRs) was first introduced in the Amazon region in 1998, with a requirement that the offset occur within the same ecosystem and with lands of greater or equal ecological value. Similar systems were implemented in the Brazilian states of Paraná and Minas Gerais around the same time. Simulations for Minas Gerais indicate that allowing such trading lowers the cost to landowners of protecting a given amount of forest. Naturally, the potential gains from trade increase with the geographic scope of trading, but this also increases the heterogeneity of incorporated forest areas (perhaps diminishing real substitutability) and the costs of monitoring and enforcement.[32]

In developed countries, TDRs have been used to control urban growth (and sprawling development), as an alternative to traditional zoning regulations. Since the 1970s, more than a hundred small-scale TDR programs have been implemented in thirty U.S. states, primarily focused on the preservation of farmland on the urban fringe. Calvert County, Maryland (a collection of communities near Washington, D.C.) adopted a TDR program

in 1978. An estimated thirteen thousand acres of farmland had been preserved by TDR sales through 2005, with some signs of influence on housing density in the region.[33]

Wetlands Mitigation Banking

Wetlands provide a rich set of ecosystem services, including water purifi-
cation, groundwater recharge, flood control, and habitat for many species of
fish, birds, and mammals. Accordingly, they present a classic public goods
problem. There are no markets for these services, and wetlands have been
rapidly depleted in many parts of the world by conversion to competing
land uses, like urbanization and agriculture. Some conversion may reflect
efficient land-use change, as wetlands give way to highly valued uses; by the
same token, in other instances the costs of losing wetlands surely outweigh
the benefits from development. The key point is that in the absence of markets for wetland services, the social value of wetlands in these transactions
is essentially ignored.

Since the early 1990s, the two federal agencies in the United States that
share responsibility for wetlands (the Army Corps of Engineers and the
EPA) have experimented with a market-based approach to this problem,
known as "mitigation banking." To secure a federal permit to convert wetlands to other land uses, a developer must compensate for the lost wetlands
by preserving, expanding, or creating wetlands elsewhere. Between 1993
and 2000, developers filled 9,500 hectares of wetlands with federal permission, and restored or created 16,500 hectares in mitigation. To meet the demand for mitigation, wetland banks have sprung up. The banks work in
many different ways, but the general idea is that developers may fill or drain
wetlands in one area, in exchange for the purchase of credits for wetlands
restoration or creation through a central broker. As of 2005, there were 405
approved wetland banks operating in the United States—nearly double the
number four years earlier. Of those 405 banks, 75 were sold out—that is,
they had exhausted their credits. Another 169 banks were awaiting approval.
Moreover, 70 percent of the operating banks were private commercial wetland banks, set up by private entrepreneurs in order to sell mitigation credits on the open market (rather than banks created to offset a specific
development). Demand for these credits appears high. In North Carolina,
for example, mitigation credits fetch thirty to nearly sixty thousand dollars
per acre; similar prices have been reported in other states.[34]

It is worth pointing out the important role played by current regulation in establishing the baseline for trading. The impetus for mitigation banking arose from a policy goal of no net loss of wetlands. In effect, mitigation banking in the U.S. is a cap-and-trade policy in which the cap on new wetlands loss (at least in principle) is zero.

Conservation Banking

A similar policy, known as "conservation banking," has been developed to mitigate the destruction of endangered species habitat. Fittingly enough, the origins of conservation banking can be traced to a trade very much in the spirit of Ronald Coase. In 1993, the Bank of America foreclosed on a parcel of land in southern California that had low value to developers, but abundant coastal sage scrub habitat—home to the coastal California gnatcatcher, a songbird that had recently been designated as threatened under the federal Endangered Species Act. Around the same time, the state's highway department, CalTrans, wanted to build a highway project through similar habitat elsewhere in the state. The U.S. Fish and Wildlife Service approved a trade: CalTrans purchased the land from Bank of America and placed a conservation easement on it, in return for permission to proceed with the highway construction. In 1995, such trades were enshrined in a state policy modeled after federal wetlands mitigation banking.[35]

The federal government's role in conservation banking stems from the underlying authority granted by the Endangered Species Act. In 2003, the program became a national one when the U.S. Fish and Wildlife Service (USFWS) issues an official regulatory guidance approving the use of conservation banking to mitigate the destruction of endangered species habitat. Much like wetlands mitigation banking, developers can purchase credits from an approved conservation bank as a means of offsetting adverse impacts on threatened endangered species. According to the USFWS, approximately forty-five conservation banks had been approved by the end of 2004, mostly in California.

Prospects and Remaining Issues

As these examples illustrate, the application of market principles to land management is an active frontier in environmental policy. Despite the excitement and activity, however, this area is fraught with potential pitfalls. One vexing issue is how credits for wetlands or endangered species habitat can be generated. For example, California's conservation banking policy grants credits not just for creating new habitat, but also for preserving existing habitat. Some critics, including the nonprofit group Environmental Defense, have argued that this approach threatens to undermine the system. If new development in one location can be offset by a promise to preserve habitat elsewhere, how do we know that anything has been gained by allowing the trade? If preservation would have occurred anyway, perhaps because it takes place on land with low value for development, then the result is a net loss of habitat relative to what would have happened without trading.

Another important drawback involves ascertaining whether land in a bank is equivalent to the land being developed. In the case of carbon dioxide emissions, or a natural resource like a fishery, equivalence is easy to determine. A pound of carbon dioxide from one automobile is equivalent to a pound from another; a boatload of fish caught by one fisher is equivalent to the same amount landed by another.

The situation is very different in the cases of natural forests, wetlands, or endangered species habitat. Land is a nonuniformly mixed resource; in plain English, not all wetlands are created equal, and this fact makes a market-based approach less appropriate. Consider a wetland in a particular location, which provides a specific portfolio of biophysical services. Many of those ecosystem services are location-specific: coastal wetlands provide nurseries for shrimp and other shellfish, for example, while wetlands farther inland help remove contaminants from freshwater flows, providing drinking water. Even if it is possible to regenerate similar services elsewhere, a different set of people will benefit. For example, some of the greatest pressures on wetlands occur in urban areas where values for other land uses are highest. We might expect, therefore, that mitigation banking would shift the balance of wetlands from urban to rural areas. If the new wetlands truly provide the same services as the old ones, that shift might be the desirable effect of reallocation by a market. But if they don't provide similar services, trading may diminish rather than increase social welfare.

One way to get around this problem might be to establish trading ratios of the sort we discussed in chapter 9. Determining the proper ratios, however, is much more difficult for land use than for air and water pollution, due to the sheer number and variety of ecosystem services and spatial concerns that must be taken into account. Moreover, measuring the ecosystem services from a particular site is a daunting problem—let alone comparing services from different sites. Another way to alleviate the problems associated with land-use trading is to impose limits on how much trading can be done. For example, developers in the United States are supposed first to avoid impacts on wetlands, and then to minimize unavoidable impacts, before they seek permission to compensate for those impacts by purchasing mitigation credits. Such restrictions on trading inevitably limit the gains from trade that can take place—a cost that should be weighed against the benefits of preventing inappropriate trades.

From an economic perspective, developers should incorporate the social costs of the damages to environmental amenities from forest, wetlands, or endangered species habitat lost to development. At the same time, there are social benefits from providing flexibility in how and where these ecosystem services are provided, in order to make room for valued development. These competing concerns suggest a potential role for market-based policies for land use; but balancing them takes care. Governments play an important role in ensuring the quality and equivalence of trades before they take place, and in monitoring and enforcing the maintenance of ecosystem services afterward.

Conclusion

These are a handful of the many creative ways in which policymakers have sought to use market principles to correct market failures—some more successfull than others. As the costs of environmental regulation and natural resource preservation become more apparent, use of these market-based approaches is likely to increase. A solid understanding of how these approaches have been implemented in the real world is therefore essential to understanding environmental policy.

We looked at the distinction between ends and means in chapter 9. Now we can see that market-based policy instruments have found their broadest application as ways of implementing environmental policy goals that have been determined without explicit regard to efficiency. To take just one example, the 10-million-ton reduction goal enshrined in the 1990 Clean Air Act Amendments was reached through political wrangling, with only a cursory reference to the marginal costs and benefits. (And as we saw, subsequent analyses of that program have found that the benefits far outweigh the costs—implying that the program, despite its ambitious scope, was not stringent enough from an efficiency perspective.)

From an economic perspective, any environmental policy that ignores efficiency is a distinctly less satisfying approach, since it gives up on the primary goal of maximizing social welfare. However, as we saw in chapters 2 and 3, the very idea of efficiency and of benefit–cost analysis stirs up considerable political controversy. It is hardly surprising, therefore, that economic analysis has exerted its strongest influence on how policies are designed, rather than on their goals.

11

Sustainability and
Economic Growth

Microeconomics, the subject of chapters 2 through 10, examines how households and firms make choices and interact at the scale of individual markets. When we ask whether the owner of a natural resource will incorporate scarcity into her extraction decisions, or whether the manager of a steel mill will take account of the damages from pollution, or how a government policy will shape the incentives of firms and individuals, we are exploring questions of microeconomics. Macroeconomics takes a more top-down view, focusing on economy-wide phenomena—the sum of millions of "micro-level" actions by households and firms. In this chapter, we address the intersection of economic growth—a macroeconomic phenomenon—and the natural environment.

You may wonder why we focus on economic growth. What is the link between growth and social welfare, the measure we have discussed throughout the text as an appropriate gauge for environmental policy choices? Economic growth has the capacity to reduce the pain of the trade-offs necessary in every social decision-making context. Enlarging the pie allows greater potential for environmental quality and other things you may care deeply about; thus, it has the potential to increase social welfare. However, just as efficiency is neither a necessary nor a sufficient criterion for sound public policy, growth is just one piece of the puzzle of social welfare. Economist Joseph Stiglitz has compared looking at economic growth to measure well-being with looking only at a firm's revenues, or a family's income, to determine well-being at an individual level. We are really interested in the balance sheet, which reports not just flows of revenues and expenditures, but stocks of assets and liabilities.[1]

In the same way that governments have the power to correct market failures and help promote efficient outcomes, governments have the capacity to spur economic growth and to help direct its course. Many of the best places to live in the world are not simply places experiencing significant economic growth, but places where income is equitably distributed, education and health care systems are good and widely accessible, and environmental quality is high. Just like households, countries may grow, but dissipate much of their income for consumption, rather than investment in these kinds of assets and institutions. Nonetheless, economic growth is a central topic in macroeconomics and the focus of much discussion among policy makers and economists. Thus it is worth focusing on growth and its implications for the environment, keeping in mind the caveats discussed above.

We begin our discussion of economic growth and the environment with the debate over the degree to which the finite availability of some natural resources can be expected to act as a limit to growth. We then define the concept of "sustainability" in economic terms, contrasting economic definitions with other common definitions. This is followed by a discussion of "green accounting," the practice of including additions to and decreases in natural capital to the calculation of traditional measures of economic growth. We end the chapter with some reflections on the viability of economic growth as a global goal, applying competing concepts of sustainability.

Limits to Growth?

In chapter 6, we solved a two-period petroleum extraction problem in which, in stylized fashion, we assessed the impact of the limited petroleum stock on efficient extraction. Our analysis suggested that private owners of nonrenewable natural resources would take the limited stock into account in determining how much to extract each year; thus the resource would be extracted efficiently. In addition, we developed the concept of marginal user cost, a measure of scarcity that takes into account economic factors as well as the physical limits of resource stocks. How are the decisions of millions of individual firms and consumers regarding nonrenewable resource extraction reflected in the global economy? The fact that many minerals that historically have been critical inputs to economic growth are available in limited quantity in the earth's crust has given rise to a contentious debate over the possibility that we will run out of some or all of those minerals, with drastic consequences for global economies.

The Limits of Assumptions

This dire scenario was the subject of the book *The Limits to Growth*, based on a systems dynamics model developed at the Massachusetts Institute of Technology in the early 1970s.[2] This large-scale computer model was designed to simulate likely future outcomes for the world economy, based on some very important assumptions. The model assumed: (1) continued exponential economic growth; (2) fixed stocks of nonrenewable resources; (3) no substitution between nonrenewables and abundant inputs; (4) no changes to the world's basic physical, economic, social, and political institutions; and (5) no technological change. The conclusions of *The Limits to Growth* and subsequent models were grim. They foresaw two possible outcomes. Either the world's nations would immediately exercise the self-restraint necessary to bring economic growth almost to a halt, thereby avoiding collision with the earth's natural limits, or the global economy would collapse within a hundred years. The collapse of the economic system would be due either to scarcity, given no new resource stocks; to pollution, should known resource stocks double and be consumed; or to population growth, should stocks double and pollution be controlled.

The Limits to Growth model of the early 1970s predicted that without immediate restraints on growth, global economies would collapse in as little as 100 years due to natural resource scarcity, pollution, and population growth. The model made very restrictive assumptions about resource scarcity, substitution, and technological and social change.

The early 1970s was a time of significant worry in the United States and other countries, given the oil shocks of that decade, high rates of inflation, rapid population growth, and increased attention to the problem of pollution in industrialized and developing countries. But the *Limits* view actually has its roots in some very old economic models. In the 1800s, Thomas Malthus, David Ricardo, and others were very concerned with the limited ability of the earth's resources (particularly land) to support fast-growing populations. And recall our discussion in chapter 6 of Stanley Jevons, the nineteenth-century economist who worried about coal depletion and hoarded writing paper, anticipating a future shortage of trees.

Cause for Optimism

Modern economists' answer to the *Limits* models has been termed "economic optimism." Like the *Limits* view, the optimistic view is represented by a particular approach, that of Julian Simon, a population economist who passed away in 1998.[3] But Simon is not the only dissenter—the *Limits* arguments are disputed by most of the discipline of economics. In the words of one analyst, the only general conclusion that one would reach from examining long-run economic growth models is that there is no general conclusion. The conclusions of such models depend critically on what they assume about some key parameters.[4] Long-run economic growth depends upon the growth of inputs, the rate and direction of technological change, and the degree to which different inputs to production can substitute for one another. Thus, the question of whether the finite availability of some inputs will slow or stop economic growth is actually an empirical question about how large these factors are in the real world. The *Limits* models make unnecessarily restrictive assumptions about these factors—assumptions that are not supported either by analysis of historical data or by general consensus regarding future trends.

The bottom line in the optimistic view is that, while resource scarcity may exert a small drag on global economic growth, the benefits of technological change have thus far outpaced the influence of these resource constraints, and will likely continue to do so. In addition, economic and political systems respond to scarcity, unlike the systems constructed by the *Limits* model. For example, scarcity causes prices to rise (as in the two-period oil extraction model we discussed in chapter 6), decreasing demand, making new natural resource stocks worth exploiting, and acting as an incentive for the development of new technologies. Increasing incomes over time tend to result in lower pollution levels, not higher levels as assumed by the *Limits* models. Income increases also tend to slow population growth, as evidenced by the very small rates of population growth—in some cases,

> *Economic models generally suggest that, while resource scarcity may exert a small drag on global economic growth, the positive effect of technological change has thus far outpaced the influence of resource constraints, and will likely continue to do so. They are "optimistic" in comparison to the Limits to Growth model.*

Growth, Income and the Environment

Several empirical analyses in the 1990s suggested that the relationship between national income and pollution or other forms of environmental degradation was hill-shaped. In other words, nations in the process of development experience worsening pollution and degradation in early stages, and then reach income levels at which these problems begin to improve. This turning point occurs, perhaps, because higher incomes increase households' preference for environmental quality relative to other goods and services, and at the same time wealthier countries are better able to afford the costs of environmental protection. Economist Simon Kuznets established such a hill-shaped relationship between economic growth and income inequality (initially increasing, then decreasing) in the 1950s; thus, this pattern with respect to pollution is often called the "Environmental Kuznets Curve" (EKC). Whether the pattern is real or not is a matter of particular importance, as some have interpreted this as a reason to suggest that economic growth will eventually repair much of the damage from early exploitation of resources like clean air and water.

Early studies confirmed this general pattern for air pollutants like particulate matter and SO_2.[5] These studies appeared prior to the development of any theoretical economic model that would explain the pattern as efficient or otherwise—the theoretical economics literature still struggles with this. In addition, the most recent work on the EKC suggests that earlier results do not hold up to better data and further scrutiny. Both the inverted U-shape and the locations of per capita income turning points established by these early studies appear to be highly sensitive to slight data variations and the specification of statistical models.[6]

While the empirical evidence for the EKC is weak, there is also no empirical support for the idea that environmental quality necessarily declines with economic growth. In fact, many studies have demonstrated in specific cases that environmental quality can improve with growth.[7] Existing data simply have not proven sufficient to map out a pollution-income relationship (if such a typical pattern even exists) over the full path of economic development.

shrinkage—observed in Western Europe and the United States, and declining rates of growth in some rapidly developing economies.

How Scarce Are Natural Resources, in Economic Terms?

In chapter 6, we discussed the fact that physical measures of natural resource reserves are insufficient economic measures of scarcity. One example of this is the petroleum reserves to use ratio—known petroleum reserves, divided by annual consumption, which presumably gives the number of years to petroleum exhaustion. What has happened to this number over time? Table 11.1 lists reserve-to-use ratios periodically from 1950 to 1990. The number does not decrease monotonically, as one would expect in a scenario in which the world was using up a finite quantity of known reserves. Instead, it increases and decreases, indicating movement along both the vertical and horizontal dimensions of the McKelvey diagram we encountered in chapter 6. In 1950, the ratio of reserves to consumption was twenty-two years—in 1990 (eighteen years after this measure, in 1950, would have suggested world petroleum resources should be exhausted) it was forty-five years.

The best way to measure scarcity from an economic perspective would be to examine trends in resource rents, which should increase as stocks dwindle. In practice, however, data on resource rents are not readily available. If resource rents are simply the difference between price and marginal extraction cost, why are they difficult to calculate? Many of the resources we have discussed in this book—clean air, clean water, some forests, and some fish—are either freely available at a price of zero, or are priced at levels that ignore scarcity rent. In these cases, prices reveal no information about scarcity. Even for natural resources for which we do believe that prices

TABLE 11.1
Reserves-to-Use Ratios for Petroleum, 1950–1990

Year	Ratio (years)
1950	22
1960	37
1972	35
1980	27
1990	45

Sources: Slade (1987), World Resources Institute (1994).

capture scarcity—privately owned minerals traded in markets—it may be difficult to separate marginal extraction cost from other costs. Finally, marginal extraction cost, in most countries, is proprietary information—data that private firms need not reveal to economic analysts. As a result, even in markets for privately owned nonrenewable resources, analyses attempting to assess economic scarcity must use indirect observation.

Attempts to assess the scarcity of nonrenewable resources through resource prices, or through scarcity rents reconstructed by statistical means, reveal that resource prices are either trendless or decreasing over time. Much has been made regarding the bet between biologist Paul Ehrlich and economist Julian Simon. In 1980, Simon challenged Ehrlich to choose a list of any five metals, worth a combined one thousand dollars. If the 1990 inflation-adjusted price of the package was higher than one thousand dollars, Ehrlich would win; if the value of the package in 1990 was lower than a thousand dollars, Simon would win. The stakes were tied to the change in the package price. Ehrlich chose copper, chrome, nickel, tin, and tungsten. In 1990, Ehrlich sent Simon a check for $576.07: The prices of all five metals had fallen dramatically over the decade.

Such tremendous price variation is reasonably common in the short and medium term. But in the long term, inflation-adjusted average prices do not appear to trend upward (as we would expect, were scarcity really a binding constraint).[8] In fact, real prices for many important nonrenewable natural resources have actually declined over time.[9]

Why might scarcity not appear to be increasing, in economic terms? After all, we are certainly consuming resources at a fast pace—in 2005, the world consumed about 84 million barrels of oil per day. Substitution possibilities and technological change provide the likely explanation.[10] In the words of one analyst, "technological progress is holding scarcity at bay."[11] Think of the many examples of this phenomenon. New seeds and chemical fertilizers have outpaced the need to cultivate marginal lands due to growing populations in many countries. Advances in finding and drilling for oil have countered the need to drill deeper and in harsher climates. Fiber optics substitute for copper as a means of information transmission. Recycled aluminum substitutes for aluminum newly processed from bauxite. Ceramics replace tungsten in cutting tools. Better irrigation technologies substitute for nonrenewable groundwater in agricultural production. Abundant sources of energy like solar and wind power, although relatively expensive today, act as ceilings on the future prices of oil, coal, and natural gas.

(After all, to the extent that renewable energy sources are a substitute for fossil fuels, no one will be willing to pay more for coal than for solar or wind.)

How Large Is the Impact of Scarcity on Economic Growth?

Even if prices are falling over time, scarcity may still dampen economic growth. For many nonrenewable resources, it may be that resource rents are rising as stocks fall, but that the contrary effects of substitution and technological change simply mask this indicator of scarcity, due to their stronger downward pressure on prices. If some inputs to important economic processes are physically limited, and those resources do not have abundant perfect substitutes, then we should be able to detect the negative impact of scarcity on economic growth through careful empirical analysis. Two analyses of note have estimated this negative impact. Nordhaus (1992) examines the influence of the limited availability of a set of nonrenewable resources, plus a set of renewable resources (including clean water and clean air) on economic growth. He estimates that the scarcity of this set of resources can be expected to slow global economic growth by a combined 0.31 percent per year between 1980 and 2050 (see table 11.2). Weitzman (1999) estimates that the limited availability of fourteen minerals important to economic growth causes a decrease in global consumption of about 1 percent per year.

How does the measured drag on growth imposed by scarcity compare to the effects of the positive force exerted by substitution and technological change? Weitzman estimates the positive welfare impacts of technological change to be about forty times the negative welfare impacts of nonrenewable natural resource depletion. Thus far, technological change has overwhelmed scarcity in the process of economic growth.

The historical evidence economists have offered cannot be considered the final word on whether economic growth at current levels is sustainable into the indefinite future. The economic argument, like arguments in other disciplines, is based on historical data and a consensus of analysts' best guesses about future scenarios. But the economic evidence overwhelmingly supports the optimistic view that nonrenewable resource scarcity will not be an important obstacle to continued economic growth in the long run.

Sustainability, In Economic Terms

We have come to the conclusion that economists are not particularly worried about running out of any specific nonrenewable natural resource. Is this reconcilable with any potential definition of sustainability?

TABLE 11.2
Estimated Drag on Economic Growth from Limited
Resources, 1980–2050

Source of Drag	Impact on world growth rate, 1980–2050 (percent)
Market goods	
Nonrenewable resources	
Energy fuels	–0.16
Nonfuel minerals	–0.03
Renewable resources	
Land	–0.05
Environmental goods	
Global warming	–0.03
Local pollutants	–0.04
Total	–0.31

Source: Adapted from Nordhaus (1992), Table 3, p. 31.

The word "sustainability" is most commonly associated with the definition offered by a report of the United Nations World Commission on Environment and Development (the Brundtland Commission, named for chair Gro Harlem Brundtland) in 1987. The commission defined the term as follows: "Sustainable development is development that meets the needs of the present without compromising the ability of future generations to meet their own needs."[12] As a goal to strive for, sustainable development by this definition is vague. Do "needs" refer to specific resources, *per se,* or just the ability to sustain specific levels of human welfare? Are the needs of future generations less than ours, the same, or greater? Are needs to be considered in absolute terms, or per capita?

Related concepts were topics of economic research well before the Brundtland Commission. But the Brundtland report prompted increased debate among economists, like analysts in many fields, to think about useful definitions of sustainability for decision-making, especially in a policy context. At the Woods Hole Oceanographic Institute in 1991, Robert Solow, the 1987 Nobel laureate in economics, offered a lecture entitled "Sustainability: An Economist's Perspective," which forms the basis of the economic definition of sustainability we will discuss here.[13]

The economic definition of sustainability requires that we leave future generations the capacity to be as well off as we are. It does not require the preservation of any particular resource; rather, it hinges on the investment of resource rents, substitution possibilities and technological change.

Sustainability, according to Solow, means leaving to future generations "the capacity to be as well off as we are today." The concept is essentially about distributional equity among generations, and it requires that we avoid "enriching ourselves by impoverishing our successors." In doing so, we must think carefully about what we use up, and what we leave behind, in an attempt to preserve an intergenerational balance sheet.

How can this be reconciled with the prospect of depleting global oil reserves, or any natural resource stock? Economic sustainability does not require the preservation of specific resources, species, things, or places. This is not to rule out the value of such preservation efforts. Even solely on the grounds of efficiency, many individual resources, species, things, or places may be worth preserving for their own sake, because their value to people is greater than the opportunity cost of preserving them. But the larger concept of sustainability, by Solow's definition, is a generalized goal not specific to any resource. Natural capital, human capital, and physical capital are all, to some degree, interchangeable. Thus, if we deplete fossil fuels to drive economic development in the twenty-first century, we must create, in the process, enough capital of other types to replace that lost value, leaving future generations the capacity to be as well off as we are today through our consumption of fossil fuels.

Sustainability, Substitution, and Technological Change

The economic definition of sustainability, like the economic arguments regarding depletion of nonrenewable natural resources, hinges on substitution possibilities and technological change. Clearly, some natural resources and environmental amenities have no substitutes. If humans value these resources highly (including both use and non-use value as defined in chapter 3), then depleting them would not coincide with Solow's economic definition of sustainability.

In most cases, however, substitution possibilities are a matter of degree. Seawater is a substitute for fresh water in industrial cooling processes, if firms invest in the materials necessary to prevent rapid corrosion of capital equipment; electric generating plants in Corpus Christi, Texas, where water has historically been in short supply, use seawater for cooling. Given a large enough investment in desalination equipment, seawater can also be a substitute for fresh drinking water; Kuwait derives 100 percent of its citizens' drinking water supply from the sea, and desalination plants have been constructed in water-scarce parts of the United States, including Tampa, Florida and Santa Barbara, California.

We are not suggesting that seawater is a perfect substitute for freshwater—it takes sea water, plus some other inputs, to achieve the quantity of fresh water desired for these different functions. Without freshwater, life on Earth would cease to exist. Obviously, a decision by world nations to consume the global supply of freshwater to fuel economic growth would not be economically sustainable. However, a decision by agricultural communities in the Great Plains of the United States to draw down nonrenewable groundwater supplies, fueling agricultural production and thereby economic development in the region, might be economically sustainable, under some conditions. In the same sense, drawing down reserves of nonrenewable oil, coal, and natural gas might be economically sustainable, under some conditions.

Substitution possibilities are, of course, easier to identify and to accept when we discuss resources like fossil fuels, water, and metals than they are when we discuss individual species, landscapes, and other natural resources and environmental amenities for which we have a particular affinity as human beings. Perhaps it is high non-use value that separates these "low-substitution" goods and services from those for which we can easily imagine substitution possibilities. Even in some of these cases, substitution may be a matter of degree. The collapse of the cod fisheries in the North Atlantic, if not reversible, is surely an ecological disaster, and regionally the cause of much economic dislocation. But while there may be no substitute for cod as a component of its resident ecosystems, populations in the United States, Canada, and Europe have substituted cod with other fish, purely for the purpose of human consumption.

Investment of Natural Resource Rents

Substitution and technological change are only part of the story when it comes to economic sustainability. In addition to substitution possibilities and technological change, economic sustainability also speaks to the role of investment of rents from the depletion of natural resources.

An example may illuminate this concept a bit. In the 1960s, oil deposits were discovered in the North Sea, off the coast of the Netherlands. The United Kingdom (UK) and Norway are the two major oil-producing countries in the North Sea. These two countries have taken very different approaches to the depletion of this valuable, nonrenewable natural resource. Rising oil prices in the 1980s made large-scale exploitation of the North Sea fields economically feasible. By the mid-1980s, the UK was producing millions of barrels of oil per day. The government of Prime Minister Margaret Thatcher supported a policy of rapid extraction, making the UK a net oil exporter, and generating substantial tax revenues used, largely, to support current consumption and to lift the country out of a long economic recession.

In contrast, exploitation by Norway has occurred at a slower pace. In 1990, Norway established a Petroleum Fund, which receives tax revenues from oil companies extracting from North Sea fields, and royalties for licenses to explore. The fund is owned by the citizens of Norway and administered by the Norwegian Central Bank. In the first quarter of 2005, the value of the fund's portfolio was approximately $170 billion.

We are not suggesting that Norway's reinvestment in non-oil capital has been sufficient to offset the loss to future generations resulting from depletion of its North Sea fields (we do not have the data to make such an assertion). But the contrast illustrates Solow's definition of sustainability; consuming the rents from resource extraction is likely not economically sustainable, but investing those rents is, at the very least, a step in the right direction. It is not depletion of natural resources, then, that damages the capacity of future generations to be as well off as we are, but rather depletion of the value of the total global capital stock—natural, physical, and human.

Problems with the Economic Definition of Sustainability

The economic concept of sustainability offered by Solow requires a good deal of knowledge about the future. To put it into practice, we must know something about the tastes and preferences of future generations, and perhaps something about the future technologies that will be available to achieve welfare gains. It is difficult to imagine how we would determine whether we are, in fact, leaving future generations the capacity to be as well off as we are without this information.

For example, over time, the tastes and preferences of residents in industrialized countries for environmental quality and natural resource amenities have increased with incomes. Should we assume that, as global incomes continue to grow, this trend will continue? In this case, future citizens will place a higher value on these goods and services than we do today, and public expenditure decisions that affect future environmental quality and natural resource management must take this into account.

Predictions about the rate and direction of technological change, even by experts, are notoriously shaky. Lord Kelvin, in 1895, is said to have claimed, "Heavier-than-air flying machines are impossible." Robert Millikan, 1923 Nobel laureate in Physics, noted in his acceptance speech that "there is no likelihood man can ever tap the power of the atom." For overestimates of the pace of technological change, one need only watch an episode of the futuristic cartoon *The Jetsons*, or the movie *2001: A Space Odyssey*, to get an inkling of just how large the margin of error is in forecasting the future.

It is not uncommon to read public documents that project the benefits and costs of various policy interventions, particularly for climate change, out to the year 2100 or further. A modicum of humility suggests that we are as blind in these predictions as analysts were in 1900 in thinking about what the world would be like today. Things change rapidly, and often in directions we cannot anticipate even a few years in advance. Constraining the direction and magnitude of those changes with respect to the goal of sustainability, based on today's myopic perspective, might result in tremendous (and unpredictable) losses in future welfare.

Given the difficulty in predicting future tastes and preferences, and the future pace and direction of technological change, it may seem that Solow's definition of sustainability is no more workable as a basis for policy decisions regarding the environment and natural resources than the Brundtland Commission definition we offered earlier, or any other. A partial response to this criticism is that we may be equally likely to err on both sides of these uncertainties about future preferences and the path of technological change. Thus, if we are acting in expectation, we may come reasonably close to the right path in managing the global capital stock with an eye to the future. Regardless of its direct workability as a basis for policy decisions, the economic definition of sustainability has much to offer in the way of insights for current decision-making.

Insights of Economic Definitions of Sustainability for Environmental Policy

Economic definitions of sustainability highlight some insights for current natural resource management and environmental protection policies. Those insights are: (1) the importance of correcting the negative externalities that arise from pollution and natural resource extraction, (2) the consideration of sustainability as a problem of inter-generational equity, and (3) the inherent conflict between inter-generational and intra-generational equity concerns.

Getting Prices Right

The first major insight we draw from the economic perspective on sustainability is that it requires dynamic efficiency—the subject of previous chapters in this book. Prices tell producers and consumers about the economic value of a good, service, or natural resource amenity—its value in use, and the opportunity cost of its consumption, including relative scarcity. So far we have framed our discussion of externalities in microeconomic terms. But their importance is magnified in the macroeconomic context. If individual firms and consumers do not bear the entire social marginal costs and benefits of production and consumption, then the aggregate consequences of their actions will diminish welfare. Without policies in place to correct environmental externalities, economic growth is unlikely to be sustainable.

Sustainability as a Problem of Inter-Generational Equity

The second major insight of economic sustainability is the importance of investment rather than consumption of resource rents. If we deplete specific natural resources in the process of economic growth, we must leave equivalent capital assets of other types to future generations so that their welfare is not diminished by this depletion.

We can think of this notion of capital investment as Pareto efficiency, with an intertemporal dimension.[14] In chapter 3, we discussed the concept of Pareto efficiency. A Pareto efficient policy is one that makes at least one party better off, and none worse off, than they were before the policy was enacted. If some parties are made worse off by the policy, then the "winners" must compensate the "losers," which is possible as long as the policy results in net gains to society. Without those income transfers, the policy might pass a benefit-cost test, but it would not be Pareto efficient. Economic sustainability is Pareto efficiency across generations. If, by applying the rules of static and dynamic efficiency that we have described in this book, we determine that depleting a natural resource stock is efficient, we must then consider whether its depletion will result in a net gain or a net loss to future generations. In the case of a net loss to the future, we are obligated to provide compensation, in the form of returns to an investment of resource rents, or some other form. Under these conditions, natural resource depletion may be economically sustainable.

On a hopeful note, the course of global economic development throughout history, while full of short-term regional ups and downs, has been quite positive from the perspective of economic sustainability. That is, past generations have, indeed, left us the capacity to be at least as well off as they were. If we consider the continually rising standards of living in most parts of the world, one could argue that past generations were "too generous" in this regard.

Conflicts between Inter-Generational and Intra-Generational Equity

Thinking about economic sustainability as inter-generational Pareto efficiency brings us to a third insight. A concern for sustainability raises the inherent conflicts between inter-generational and intra-generational equity. If sustainability is about equity among generations, can it tell us anything about equity within generations? Schelling (1997) pointed out the paradox of considering the expenditure of large sums today to "purchase" a climate less influenced by anthropogenic carbon dioxide emissions tomorrow, when those funds could be used to purchase increased quality of life for today's poor.

This is a particularly stark point, given that most future beneficiaries of actions taken today to prevent global climate change will reside in developing countries, and that future residents of these countries will almost certainly be wealthier than today's residents. Such policies, then, may amount to billing poor residents of developing countries today for welfare improvements accruing to wealthier residents of these countries tomorrow. Note that this paradox does not require that developing countries actually pay out of pocket for climate change policies today—only that they pay in terms of reduced resources available for mitigating the impacts of poverty and environmental degradation today. One dollar spent on improving environmental quality and natural resource amenities for future generations is one dollar not spent on improving welfare (environmental or otherwise) today.

Keeping Track: Green Accounting

A consideration of economic sustainability forces us to think more carefully about whether or not we are acting as good stewards of the world's capital stock. If sustainability requires that we consume and invest such that future generations have the capacity to achieve our own level of well-being, it also requires some method of keeping track of whether or not this preservation of the global capital stock is actually occurring. Adjusting standard indicators of economic growth to reflect natural resource stocks and the state of the environment is an important step in keeping track of our progress toward that goal.

Traditional indicators of economic growth are measures of goods and services produced by labor and property either within a country (gross domestic product, GDP) or supplied by a country's nationals (gross national product, GNP). These gross measures can also be converted to net measures by subtracting capital depreciation, creating net domestic product (NDP) and net national product (NNP).[15]

Each of these traditional measures of economic growth excludes non-market activities, such as household production (cooking, cleaning, child care, and home improvements, for example), and black markets for goods and services. Most importantly for our purposes, they also exclude environmental services like clean air and water; and the value of natural resource stocks, such as oil, coal, forests, and fish. Many economists have suggested that a true measure of economic growth should account for changes in the value of these assets. A measure of growth that includes the depreciation of not only the stock of physical capital, but that of natural capital, as well, is often called "green NNP".

Should Traditional Measures Be Changed?

The debate over whether the traditional measures of economic growth were sufficiently inclusive began well before the phrase "sustainable development" was first uttered. The fathers of national income accounting in the United States, the process that undergirds the calculation of growth measures, were well aware of its shortcomings. Simon Kuznets, who received a Nobel Prize for his work in developing the U.S. national income and product accounts, did not intend these growth measures to become measures of social welfare, as they have often been used. Income accounting was designed with many important goals in mind, including the provision of indicators of an economy's performance over time, measurement of savings and investment, and tracking of business cycles.

The information produced by this process is invaluable to governments seeking to implement policies to promote or control the pace and direction of economic growth. Nonetheless, A. C. Pigou noted, with irony, that "If a man marries his housekeeper or his cook, the national income is diminished." More to the point, he states:

> It is a paradox, lastly, that the frequent desecration of natural beauty through the hunt for coal or gold, or through the more blatant forms of commercial advertisement, must, on our definition, leave the national dividend intact, though, if it had been practicable, as it is in some exceptional circumstances, to make a charge for viewing scenery, it would not have done so.[16]

A measure of growth that includes the depreciation of not only the stock of physical capital, but that of natural capital, as well, is often called "green NNP".

The paradox identified by Pigou has wide-ranging implications. If the owner of an oil well in Texas pumps out her remaining reserves and sells the oil, the value of her sale is added to U.S. NNP, the depreciation of her well pump is subtracted from U.S. NNP, but nothing is recorded to account for the value of the oil no longer beneath the ground in Texas. The mining of one ton of coal from a U.S. mine increases GNP by $17; subtracting the depreciation of mining equipment and the one-ton decrease in coal reserves would bring that number down to about $5.50; the number would be even lower were we to consider the pollution externalities from mining and burning coal.[17] The value of privately owned domestic livestock lost to disease is deducted from traditional measures of NNP; commercial fisheries depletion is not.

We have spent a good deal of time in this book describing how economics treats natural resources as capital assets in determining efficient extraction rates, but the exclusion of these assets from measures of economic growth is inconsistent with this practice. Even where marketed natural resource commodities make important contributions to national output—oil, fish, and timber are good examples, particularly in some developing countries—additions to and subtractions from the stocks of these resources are not included in the calculation of NNP.

If these shortcomings of traditional measures were recognized from the outset, why has their calculation not been adjusted to account for these excluded portions of economic activity? There are two main reasons why this is the case. First, the values of nonmarket goods and services are, by definition, difficult to capture. Even for marketed goods and services that are currently excluded (such as minerals, fish and timber), the ease of measuring changes in stocks varies greatly. Most of the initial experiments with green accounting that we will discuss in the next section have started with the easiest categories, minerals and timber.

Second, as a measure of economic growth, traditional NNP is likely to be strongly correlated with a true, all-inclusive measure of NNP. However, if the economic value of excluded goods and services has increased over time, the inherent bias in using traditional NNP as a proxy for true NNP has likely grown, as well. For example, the value to the U.S. population of recreational opportunities in forests and other undeveloped lands has increased markedly. These uses, in some cases, conflict directly with commercial use of such lands. Commercial timber extraction and mining generates additions to NNP, but no subtractions are made to represent the opportunity cost of potential recreation. Over time, this omission has grown in importance, due to the growth in the value of recreational activities, making NNP an increasingly biased measure of U.S. economic activity.

Experiments with Green Accounting

Some of the experiments with greening the national income accounts have taken place in industrialized countries, where the idea began to take root in the 1970s. Two economists at Yale asked the question, "Is Growth Obsolete?" and attempted to incorporate things like traffic congestion, crime, and, to some extent, natural resource depletion, into U.S. economic growth measures.[18] Their comprehensive welfare measure increased by about 42 percent between 1929 and 1965, less than one-half of the increase in traditional per capita growth measures over this time, but still a substantial increase. Around the time that Norway began to exploit its North Sea oil deposits, Norwegian analysts focused on the issue of how to account for the depletion of this resource in calculating national economic growth estimates. Norway has since estimated green NNP, accounting for depletion of oil, fisheries, forests, and even clean air (by including some air pollutant emissions).

The 1980s also saw significant attention paid by analysts in industrialized countries to natural resource depletion in the developing world. This was a time of particular worldwide public concern over issues like deforestation, particularly in the Amazon basin. A book called *Wasting Assets* estimated that the phenomenal rates of economic growth posted by Indonesia in the 1970s and early 1980s would have been halved had it taken into account the depletion of timber, oil, and other resources.[19] In the aftermath of this work, many developing countries, themselves, became interested in adjusting NNP to account for natural resource depletion. Experiments are ongoing in the Philippines, Namibia, and other developing economies.

A relative latecomer to this process, the United States produced its first official effort at incorporating resource depletion into the national income and product accounts in 1994. This initial effort included only selected mineral commodities, including oil, gas, and coal. Opposition from the U.S. Congress resulted in the suspension of further efforts on green accounting, despite a highly favorable report by a National Research Council external review.[20] The accompanying box describes a more comprehensive effort to understand the possible differences between traditional economic growth measures and more comprehensive ones.

General Conclusions from Theory and Experience

There are some general conclusions to be drawn from past and current experience with greening the national income and product accounts. First, the difference between green NNP and traditional NNP tends to vary by country and over time. An economy's relatively greater dependence on resource extraction tends to create a relatively greater difference between the two measures. This is especially true of green NNP calculations that include natural resource extraction, but exclude environmental externalities like air and water pollution. For example, in the calculations of NNP that were done for the United States in 1994, subsurface minerals depletion appears to have been approximately counterbalanced by exploration and discovery of new resources. In contrast, in the late 1980s, the green NNP calculated for Indonesia, in *Wasting Assets,* departed quite drastically from measures of traditional NNP. The picture for the United States might be quite different during an earlier period of its development, when natural resource extraction comprised a larger share of total economic activity.

Second, traditional NNP not only leaves out natural resource depletion and environmental externalities, but it also omits the contribution to NNP of technological progress. Rough calculations indicate that including technological progress in the calculation of NNP would result in a substantial upward correction, perhaps by as much as 40 percent.[21]

GDP versus "Genuine Wealth"

A group of economists and ecologists have compared actual rates of change in traditional GDP, 1970 through 2000, to those of their own estimated measure of "genuine wealth," which incorporates some natural resource depletion, population growth, and a proxy for changes in technology.[22] Rates of change in estimated per capita genuine wealth over this period in the United States, Bangladesh, India, Nepal, and Pakistan are positive, but significantly lower than the growth rate of per capita GDP. For the United Kingdom, the growth rate of estimated per capita wealth is approximately the same as that of traditionally measured per capita GDP. In China, the more comprehensive wealth measure has grown at a rate substantially higher than traditional GDP. In sub-Saharan Africa and the oil-exporting Middle East/North Africa region, the more comprehensive wealth measure actually decreases between 1970 and 2000.

The analysts point out that concluding from this analysis that the poor countries, particularly in sub-Saharan Africa, are consuming too much and that the rich countries, like the U.K. and the U.S., are following a more sustainable path would be a mistake. In fact, the poor countries suffer from too little investment, as well as too little consumption; the rich countries may, in contrast, be succeeding by this measure through the import of natural resources and resource-intensive products from the poor countries (thereby avoiding depleting their own resources).

Including both resource depletion and technological progress in national income and product accounting, which compares national economic conditions across countries, as well as determines the direction and pace of global economic growth, would be an "almost practical step toward sustainability."[23] However, searching for a way to keep track of what is happening to the world's natural capital within individual nations' annual balance sheets will leave out changes in some very important assets: namely, global commons, like oceans, stratospheric ozone, and the upper atmosphere (currently a major receptacle for the world's greenhouse gas emissions). In addition, estimation of green NNP on the national scale masks the depletion of natural resources through trade; resource-adjusted NNP figures for poor countries that deplete forests or minerals to fuel economic growth will be diminished due to this practice, while the same figures for rich countries

Trade, Growth, and the Environment

The removal of barriers to trade was one of the hallmarks of the late twentieth century. Regional agreements like the North American Free Trade Agreement (NAFTA) took shape, along with further evolution of the Global Agreement on Tariffs and Trade (GATT), and then the World Trade Organization (WTO). Open economies generate higher levels of social welfare—this is an important economic principle. How does this macroeconomic phenomenon interact with natural systems?

In many cases, trade agreements are concluded by coalitions of countries among which environmental regulation and natural resource management policies vary significantly. Economic theory suggests that stringent efforts by some countries to internalize the costs of pollution through environmental regulation will alter international trade patterns, resulting in the export of dirty industries to countries with less stringent regulations. This has been called the "pollution havens hypothesis." The underlying theory and intuition behind this concept are strong, yet, through the mid-1990s, studies found no evidence for the impacts of environmental regulations on trade patterns.[24]

Statistical tests of this hypothesis are complicated by two phenomena—unobservable heterogeneity and endogeneity.[25] By the first term, we mean that unobserved industry and country traits can be correlated with the likelihood of regulation and the export of pollution-intensive goods. This poses the danger of seeing in the data a causal link between trade and pollution where none really exists. By endogeneity, we mean that the direction of causation between trade, regulation, and pollution is unclear. Environmental regulation often appears in countries with substantial international trade. Does trade influence pollution regulation? If so, this is the opposite effect suggested by the pollution havens hypothesis, which posits that regulation influences trade.

Recent studies accounting for these statistical challenges have established pollution haven effects where trade is between high- and low-standard countries and industries are mobile.[26] Small effects have also been measured in trade among U.S. states.[27] Another recent study finds no evidence of trade's detrimental environmental effects, finding instead that trade tends to reduce concentrations of local air pollutants.[28]

that import these products will not. An accompanying box describes the state of the evidence regarding the effects of trade on the geographic distribution of pollution and resource degradation. Estimates of *global* green NNP, taken regularly, would be necessary to illuminate the trade-offs taking place among natural and other assets in the pursuit of economic growth.

Are Economic Growth and Sustainability Compatible?

Among the competing definitions of sustainability are many that conflict with the economic definition that we offered earlier. In fact, the economic definition has been termed "weak sustainability," in contrast to "strong sustainability," which adds the additional requirement that natural resource stocks not be further depleted nor environmental quality further degraded in pursuit of economic growth. This constraint amounts to a restricted view of substitutability among natural and physical or human capital.

In their concept of sustainability, economists see substitution and technological change as the rule, and limits to these forces in supporting economic growth as the exception. An alternative view, a perspective often labeled "ecological economics," holds binding scarcity as the rule, and sees the ability of substitution and technological change to overwhelm scarcity in pursuit of economic growth as the exception—something possible, perhaps, over the first few millennia of human development, but not indefinitely.[29] Adherents of this view see serious conflicts between continued economic growth and sustainability. Ecological economists worry that the pursuit of economic growth and development for their own sake overlook the critical importance of the appropriate scale of economic activity.

These are two very different worldviews. One may subscribe to either of these views on sustainability, and still draw important lessons from the economic definition of sustainability offered in this chapter. For example, getting prices right, justified on efficiency grounds alone, is also a step in the direction toward either of these concepts of sustainability—one that sees immovable limits to growth everywhere, and one that acknowledges few insurmountable limits. The issues of inter-generational and intra-generational equity we raise (such as the contrast between consuming and investing resource rents) are also relevant to either view of sustainability.

Conclusions

In this chapter, we have explored the links between the environment and the macroeconomic phenomenon of economic growth. We began with a big-picture look at the issue of nonrenewable resource scarcity that we approached from the level of individual firms in chapter 6. Some have postulated that the physically limited stocks of non-renewable resources like oil and copper will eventually act as serious brakes on economic growth. We examined the historic evidence for the negative impact of scarcity on economic growth, and found that this impact, while measurable, is much smaller than the positive impact of substitution and technological change.

We developed an economic definition of sustainability, which requires leaving "the world the capacity to be as well off as we are today." The definition hinges upon substitution possibilities and technological change. While there are some serious complications in applying this rule of thumb directly, it offers significant insights for public policy regarding the environment and economic growth. First, it offers additional support for getting prices right, and sorting out the market failures initially discussed in chapter 5—this is a step in the right direction, whether one adheres to the economic definition of sustainability, or definitions that encompass stronger conservation principles with respect to specific natural resources. The economic definition of sustainability also speaks for investment over consumption of the rents from natural resource extraction, and it prompts us to think more clearly about the trade-offs between achieving goals of intra-generational and inter-generational equity.

Since the creation of the practice of national income accounting, economists have worried that growth measures like the rate of change in net national product omit many portions of the economy, including natural resource depletion and environmental degradation. We determined in this chapter that more comprehensive measures are needed, particularly if we are to apply the economic concept of sustainability, which forbids us from depleting the total global capital stock, including natural, physical and human capital.

While economic growth is not synonymous with growth in social welfare, it does have the capacity to increase welfare, making somewhat less painful the necessary trade-offs in "sacrificing some of one good thing for more of another." For example, as incomes rise, countries are able to afford improvements in health, education, and environmental quality—whether they choose to invest in these, or in something else, can have important consequences for social welfare.

12

Conclusion

We have now come to the end of our journey. Rather than try to summarize all that we have discussed, we will consider some of the broader implications of economic analysis for environmental policy.

What Does Economics Imply for Environmental Policy?

In July 2002, *The Economist* published a story entitled "The Invisible Green Hand." One of the pull quotes in the story was: "Markets could be a potent force for greenery—if only greens could learn to love them."[1]

Sometimes the quote rings true—perhaps more often than you thought, if this was your first exposure to environmental and resource economics. As we have seen throughout this book, economic theory provides strong arguments in support of active (if enlightened) government policies. For the most part, economic analysis would suggest that environmental amenities are underpriced, and that renewable natural resources are overexploited. For example, subsidies for fishing or the extraction of timber from public lands are inefficient, in part because they result in environmental losses. Studies of the U.S. sulfur dioxide trading market have found that benefits vastly exceed costs, implying—if anything—that regulation ought to be *more* rather than less stringent.

In other cases, of course, economic analysis arrives at conclusions that an environmental advocate might object to. For example, benefit-cost analyses of the U.S. Clean Water Act suggest that the act's original goal of zero emissions to all U.S. water bodies, its focus on regulating point-source pollution, and its primary reliance on technology standards have resulted in a federal regulation with substantial net costs. Many economists would argue that

suburban sprawl, despite its aesthetic shortcomings, may in fact be the socially desirable reflection of individuals' willingness to pay for certain amenities (like quiet streets and large yards).

The underlying point is that environmental economics is not "green"
or "brown," but neutral. Economists treat natural resources and environmental amenities, in effect, like any other assets. On one hand, this approach
brings environmental services onto the balance sheet, ensuring that they are
not given a value of zero in public debates over whether to extract oil from
pristine wilderness areas, or whether to allow the construction of new housing in a forested or agricultural urban fringe. On the other hand, treating
natural resources and environmental amenities in the same framework as
other goods and services reflects an underlying assumption that substitution
possibilities are just as relevant to environmental and natural resources as
they are to other assets. To be sure, there may be many specific resources
(like the Grand Canyon, or portions of the Amazon rainforest) that should
be preserved for their own sake even on efficiency grounds. But in many
other cases it will be efficient to convert forested lands to urban or agricultural use, to deplete a nonrenewable groundwater aquifer, or to tolerate
some amount of pollution in exchange for the services provided by a polluting industry. Efficiency and cost-effectiveness analysis in all of these cases
should serve as a starting point for discussion regarding environmental and
natural resource management policies—not the final word.

The Roles of Firms, Consumers, and Governments

We read a lot these days about the power of consumers to affect the practices of firms whose activities may result in pollution or other natural resource damages, about "corporate social responsibility" and its implied
voluntary measures to reduce the environmental impacts of economic activity. There are certainly instances in which firms improve their environmental performance beyond what is required by law, and cases in which
consumer pressure has resulted in substantial improvements in environmental and natural resource outcomes. Nonprofit environmental advocacy organizations also play an important role in this process.

However, an important message of this book is that as long as the markets for environmental amenities are incomplete, consumer pressure and
voluntary efforts by companies to reduce their impacts on the environment
will not be sufficient to achieve the efficient level of pollution control, or
efficient natural resource management practices. The incentive structure

created by the "wrong prices"—prices that do not reflect the full social cost of engaging in an environmentally damaging activity—is simply too powerful.

For example, hundreds of thousands of Americans may be Sierra Club members, but does each member contribute annually an amount equal to their true willingness to pay for wilderness preservation and the organization's other ambitious goals? Given our discussion of public goods and free riding, you should suspect that the answer is "no," and additionally that there are likely many nonmembers of the Sierra Club who have some willingness to pay for the services it provides. Similarly, many firms trumpet their activities in environmental stewardship, and many of these efforts are sincere and result in the generation of substantial environmental benefits. But we cannot rely on these voluntary activities, alone, to correct the substantial market failures that contribute to inefficient environmental degradation and resource depletion.

Where markets are absent or otherwise incomplete, well-designed public policies, like market-based instruments for pollution control, are needed to correct these incentives and get the prices right. These policy instruments use market principles to correct market failures, and align private incentives with public ones. This is not to say that government regulation is always the answer to market failure in the environmental realm. In fact, in this volume we have noted a number of examples of government failure—the role of government in worsening environmental outcomes by subsidizing resource extraction, for example. But from the economic perspective, government is a necessary central authority with the power to tax citizens for the provision of public goods, implement and enforce regulations that internalize the external costs of pollution, and create and enforce property rights structures that promote resource stewardship, rather than excessive depletion. Without this, the voluntary actions of firms and active participation of citizens can ameliorate the impact of incomplete markets on environmental degradation, but they cannot eliminate it.

Some Final Thoughts

Good economic analysis formalizes and makes transparent the difficult compromises inherent in decisions regarding the use and management of natural resources and environmental amenities. Applying economic principles to environmental policy choices comes as naturally to economists as doing so in choices about other aspects of the economy. The "environment" is not separate from the "economy" in the framework we have offered in this text—indeed, environmental problems cannot be fully understood without a basic intuition for how markets function, and how they fail.

Correcting market failures in this realm, as in any other, is efficient. It is good for the economy due in part to the benefits generated by the resulting environmental improvement. This is not to say that there are no trade-offs to be made. Internalizing the pollution externality of SO_2 emissions from power plants may force some high-sulfur coal miners out of their jobs, resulting in substantial losses for coal-mining families and communities. Establishing property rights over open-access fisheries will drive some high-cost fishers out of fishing altogether. We have discussed many such examples in this text. Market-based approaches to correcting market failures can reduce the total costs of environmental regulation and efficient natural resource management, but they cannot eliminate them.

Economic analysis, combined with careful consideration of equity issues, shrewd political strategy, and other inputs, will help students of environmental studies to make better decisions about environmental policy, and to better interpret the consequences of others' decisions. While efficiency and cost-effectiveness are not the only criteria that contribute to sound environmental policies, they can help us make conscious choices about how much of one good thing must be sacrificed to have more of another, just as we do in daily decisions about our own household budgets. In this text, we have shown that economics can make vital contributions to both the analysis of environmental problems and the design of possible solutions. We hope that the tools we have introduced help to illuminate the environmental issues you will approach throughout your coursework and your career.

Discussion Questions

Chapter 2

1. Explain why it would not be desirable (from the point of view of economic efficiency) to eliminate all emissions of sulfur dioxide from electric power plants. Now suggest a setting in which zero pollution might be efficient.

2. In figure 2.3, abating X^{MAX} units of pollution would result in total benefits of pollution abatement greater than the total costs. Is this amount of pollution abatement efficient? Why or why not? Refer to both figure 2.3 and figure 2.6 in your answer.

3. Consider the case of Aracruz Celulose, S.A., the pulp manufacturer described in this chapter. Suppose that a study finds that the marginal benefit of reducing chlorinated organic compounds in the effluent from pulp mills is twenty-five dollars per kilogram of AOX. (For simplicity, assume that this number does not change with the amount of pollution.) Which pulping technology would be efficient in this case? What would be the resulting annual costs and benefits associated with pollution abatement?

4. Explain the typical shapes of the total abatement cost and benefit curves, as well as the marginal benefit and cost curves.

5. Imagine a policy to reduce emissions of greenhouse gases that would yield a marginal net benefit, in present value, equal to $1 million today and $2 million next year. Would this policy be dynamically efficient? Why or why not?

6. Advocates for wilderness protection have often criticized the common practice of setting aside "rock and ice" (i.e., alpine areas) as protected land, while more productive bottomland areas (along river valleys, for example) are typically left in agriculture or other intensive use. In general, the bottomland areas would also provide richer species habitat than the alpine areas. What are the likely pros and cons of this approach from the perspective of economic efficiency?

Chapter 3

1. How might we estimate the marginal benefits of preservation of the California condor, the endangered species discussed in the first box in chapter 3? Imagine that we estimate the marginal benefits of condor preservation to be two hundred thousand dollars. Are there policy measures that would be excluded if we applied the equimarginal principle to condor preservation? Explain.

2. The economic benefits of an environmental policy, like the reduction of sulfur dioxide emissions from power plants, are measured by the collective willingness to pay of human beings. Policies like this may have ecological benefits, like the effects of reductions in acid rain. Discuss the degree to which measuring the economic benefits of a pollution-reduction policy will capture ecological benefits.

3. Why do economists generally prefer revealed preference approaches to environmental benefit valuation over stated preference approaches? Are there cases in which stated preference approaches would be recommended?

4. Explain the difference between use and non-use value, with reference to a particular environmental policy in which you may be interested, like greenhouse gas emissions reductions or endangered species preservation.

5. Should environmental and natural resource management policies be put to a strict benefit-cost test? Why or why not?

6. Some environmental laws in the United States explicitly prohibit the use of benefit-cost analysis in some areas of environmental policy. For example, the Clean Air Act declares that air quality standards are to be determined purely on the basis of protecting public health with "an adequate margin of safety," and forbids the administrator of the Environmental Protection Agency from considering costs in setting stan-

dards. Can you provide a critique of such an approach, from the perspective of economic efficiency? What might be the consequences of such an approach? Now take a step back. From your own perspective, do you think such an approach is advisable? Why or why not?

7. Contrast the economic perspective on endangered species preservation with the perspective that individual species have infinite value. What are the implications of each perspective for public policy?

8. Economic analyses indicate that reducing timber extraction in the U.S. Pacific Northwest, in an effort to preserve old-growth habitat for the endangered Northern spotted owl, had significant net benefits. Yet many logging communities experienced significant economic dislocation as a result of these policies. Discuss the links, if any, between economic efficiency and distributional equity in this case. Was reducing timber extraction a Pareto improvement?

Chapter 4

1. Explain why a demand curve can be considered to be a marginal benefits curve, and why a supply curve is equivalent to a marginal cost curve. Refer to a specific environmental problem in your answer.

2. Describe a situation in which you would expect the free market to result in an efficient outcome with respect to environmental quality or natural resource management, and one in which you would not.

Chapter 5

1. Describe the efficiency loss that results when the social costs of pollution are external to the private costs of producing a good like electricity. How do the market price and quantity of electricity with pollution externalities compare to the efficient price and quantity? What are the potential gains to society from regulating pollution?

2. Examine figure 5.1, and imagine that the marginal damages of pollution were much "flatter" than the curve represented here. How would this change the magnitude of the deadweight loss from pollution in this case? What are the intuitive implications?

3. Contrast a pure public good, like biodiversity preservation, with an open-access resource, like some fisheries and groundwater aquifers. How do these classes of goods and services differ, and what are the implications for environmental policy?

4. Many people contribute money to environmental advocacy organizations. Can we measure the benefits of the services provided by these organizations by summing up contributions? Why or why not?

5. Explain the link between public goods and positive externalities.

6. In the international environmental treaty "game" described by figure 5.5, both countries would be better off if each contributed to the clean-up of a shared pollution problem, but this is not what we expect to happen. Why is this better outcome unlikely to occur?

Chapter 6

1. If we take the world's known reserves of oil, and divide this total quantity by average annual oil consumption, we obtain the "reserves-to-use ratio," the number of years that remain before exhaustion of our oil resources. Explain why this ratio paints a misleading picture of oil scarcity.

2. The Hotelling Rule states that marginal user cost rises at the rate of interest. Explain the intuition behind this result.

3. Under what conditions would you expect the extraction rate of a nonrenewable natural resource to depart from the dynamically efficient rate?

4. Oil and endangered species are both natural resources with high economic value. Yet a private landowner in the United States might react very differently to the discovery of an oil well on her property than she would to the discovery of an endangered species population. Explain this difference, using economic concepts.

5. Economist Robert Solow has said that "the monopolist is the conservationist's friend." Explain this in the context of nonrenewable resource extraction.

Chapter 7

1. The models we explore in chapters 6 and 7 (nonrenewable resources, fisheries, and forests) have all referred to the concept of economic rent. Define rent, and explain its relationship to economic efficiency.

2. Explain the relationship between the biological timber rotation, the Wicksell rotation, and the Faustmann rotation. Which one is economically efficient, and why?

3. Forests generate nontimber benefits. Explain the effects of the following nontimber values on the efficient timber rotation: (1) new growth

provides habitat for white-tailed deer, (2) old growth forest provides habitat for the endangered red-cockaded woodpecker, and (3) decaying trees provide habitat for an important species of insect.

4. Explain the link between the establishment of property rights and deforestation in tropical countries. Given our discussion in this chapter, does this imply that large-scale privatization of land in tropical regions is a solution to deforestation?

5. A fishery will always be economically overfished before it is biologically overfished, thus the efficient level of fishing effort is lower than the biologically efficient level. Explain why this is true.

6. Under open access, fishing will occur until the total benefits are exactly equal to total costs, and net benefits are equal to zero. Why don't fishers stop entering the fishery before this happens?

7. Explain precisely why reducing fishing effort from the open-access equilibrium would be efficient. What would be the effect on fishing communities in the short run? In the long run?

8. Imagine that a national government reduces the marginal cost of fishing effort through a subsidy, such as a fuel tax exemption. How would this subsidy affect the open-access equilibrium level of fishing effort, represented in figure 7.6? Would the net benefits of the fishery still be equal to zero in this case?

9. History offers many examples of small groups of farmers collectively managing shared irrigation systems, yet there are also many examples of inefficient depletion of large-scale groundwater aquifers due to agricultural irrigation. Use the concepts of common property and open access to describe a potential explanation for these two very different phenomena.

Chapter 8

1. The Coase Theorem suggests that, under some conditions, private bargaining will resolve negative externalities. Does this mean that pollution problems should be left for the market to solve? Why or why not? Discuss this from the perspective of efficiency, as well as equity.

2. According to Ian Parry and Kenneth Small, the efficient gasoline tax in the United States is somewhat less than 1 dollar a gallon (see text box). Of this amount, only six cents is related to global climate change from carbon dioxide emissions. Some people might argue that six cents is far too little to change people's behavior, for instance by in-

ducing them to drive less or buy more fuel-efficient cars (and even a dollar might not make much difference). If the Pigouvian tax turns out not to make much difference in what people do, is it still efficient? Why or why not?

3. Why has the price of carbon dioxide emissions credits on the voluntary Chicago Climate Exchange been so much lower than the price of carbon dioxide emissions credits on the European Union's Emissions Trading System? What general conclusions can you draw from these contrasting examples?

4. How would a landing tax in a fishery work to reduce the amount of fish caught?

5. Many economists have concluded that the marginal benefits associated with reducing greenhouse gas emissions are flat, since each additional ton of carbon dioxide (or its equivalent) has the same effect on global warming. At the same time, there is considerable uncertainty about the future marginal costs of controlling greenhouse gases. What do these two assertions imply about the choice between a tax and a tradeable permit system, on the grounds of efficiency?

6. Much of the political debate surrounding the use of emissions trading concerns the allocation of the pollution allowances—in particular, whether to auction off the allowances, or give them away for free. Explain why the method of allocation does not affect how much pollution firms end up controlling, assuming that transactions costs are low. What do you think would happen if transactions costs were high?

Chapter 9

1. Explain the intuition behind why market-based instruments (emissions taxes and tradeable permits) are cost-effective, while uniform standards are generally not.

2. Redraw figure 9.1 using an even flatter marginal abatement cost curve for firm A and a steeper one for firm B. What would happen to the cost-effective allocation? What would happen to the size of the cost savings from a market-based instrument, relative to a uniform standard? Explain.

3. The 1977 Clean Air Act Amendments required new electric power plants to install scrubbers in order to remove sulfur dioxide emissions. Such an approach is often called a "technology-forcing" approach, and is typically promoted as a way of ensuring that polluters install the

most advanced or best available abatement technology. From an economic perspective, what kind of incentive does such a policy provide for the development and adoption of new technologies?

4. An increasingly common policy prescription for reducing greenhouse gas emissions, often called the safety valve approach, proposes that governments set a cap on the price of pollution allowances. For example, a government establishing an emissions trading system for carbon dioxide might offer to sell an unlimited amount of permits at a price of one hundred dollars per ton, which would effectively prevent the market price from rising higher than that level. A criticism of such an approach is that it would undermine incentives to develop and adopt new pollution abatement technologies. Why is this the case? Do you agree with the criticism? Why or why not?

5. Why are hot spots a potential problem with market-based instruments? Why would location-specific taxes (or trading ratios) help to alleviate the problem?

Chapter 10

1. Why is the U.S. sulfur dioxide allowance trading program widely considered to be a success? In your answer, be sure to discuss the policy's environmental performance, cost-effectiveness in comparison to other potential policies, compliance and enforcement, and distributional implications.

2. One of the strongest objections to market-based fishery management is the possibility of "consolidation." Describe this phenomenon and explain why it may occur under an individual tradable quota (ITQ) system. To what degree has consolidation occurred in New Zealand fisheries managed by ITQs? Is it a concern for efficiency, distributional equity, or both? The government of New Zealand has taken some measures to reduce the impact of consolidation. Describe these measures, and discuss the equity/efficiency trade-off that they imply.

3. Draw an analogy between the U.S. sulfur dioxide allowance trading program and a hypothetical market for water consumption permits during a drought. How would these two policies be similar? How would they differ? Be sure to address the source of the potential gains from a market-based approach in each of these cases.

4. Numerous successful tradeable permit systems for air pollution control have emerged in the United States in the past two decades, but

experiments with water quality permit trading have been much less successful. Offer some potential reasons for this difference.

5. Is a "pay as you throw" policy for solid waste management a Pigouvian tax? Why or why not? Have these policies been successful?

6. What are some of the barriers to large-scale application of market-based policy approaches to land management? To what degree can these barriers be overcome?

Chapter 11

1. The *Limits to Growth* model and economic models differ significantly in their assumptions. Describe these different assumptions, and the resulting differences in what the models suggest regarding the limits on future economic growth posed by the finite availability of important resources like oil and coal.

2. The world's supply of oil is being depleted much faster than the rate of natural regeneration. From an economic perspective, can this be efficient? Can it be sustainable?

3. Economist Robert Solow describes green accounting as an "almost practical step toward sustainability." Why is it *almost* practical? How might green accounting promote sustainability from an economic perspective?

4. If we were to try to implement the economic concept of sustainability, we would face some important sources of uncertainty. Describe these areas of uncertainty, and how they might limit our ability to implement sustainable policies.

5. What are the most important insights of economic sustainability for current policies regarding natural resources and the environment?

6. In 1976, the state of Alaska established the "Alaska Permanent Fund," which primarily uses the returns from investing the proceeds of the sale of oil to provide Alaska residents with dividends (averaging $1,240 in the past fifteen years). The Permanent University Fund in Texas (currently valued at $15 billion) uses proceeds from the sale of oil leases and royalties on state land to finance several state universities. Assess these policies from the perspective of economic sustainability.

References

Chapter 1: Introduction

1. Intergovernmental Panel on Climate Change, Working Group II (2001), *Climate Change 2001: Impacts, Adaptation, and Vulnerability: Summary for Policymakers*, 2.

Chapter 2: Economic Efficiency and Environmental Protection

1. We have given a somewhat loose explanation for why marginal cost equals the slope of the total cost function. A precise explanation requires calculus: Marginal cost is (by definition) the first derivative of the cost function, which in turn is the slope of the total cost function. But even if you haven't had calculus, we can make the discussion a bit more rigorous. For any given change in abatement, the change in total cost is the same as the change in the height of the cost function. To see this, refer back to the cost function in figure 2.1. Choose two nearby levels of abatement, which we shall call point A and point B. Suppose we increase abatement of some pollutant from A to B. The change in total cost that results is obviously the total cost at B minus the total cost at A. For such large changes, we need to take into account the curvature of the cost function. As the distance between A and B gets smaller, however, the curvature matters less and less. Indeed, for very small changes in abatement the cost function is almost a straight line, and the change in height between A and a nearby point can be measured by the slope of the function at A times the distance between the points. In particular, the change in total cost between A and $A + 1$ (assuming that we are measuring units so that one unit is sufficiently small) equals the slope of the function at A. But we defined marginal cost as the change in cost from one more unit of abatement. Thus the marginal cost at point A is just the slope of the total cost function at that point.

2. In fact, summing up the areas of these rectangles would only give an approximation of the actual area under the curve. But as we made the rectangles narrower and narrower (and drew more and more of them to fill in abatement from zero to X_L) the total area of those rectangles would be a better and better approximation to the area under the curve. Eventually, as the rectangles became increasingly narrow—"in the limit," as mathematicians like to say—this approximation would become perfect.

3. To be more precise, if the efficient or optimal policy in one period is independent of the policies in all other periods, the problem to be solved is a static efficiency problem. If optimal policies are correlated across periods, the efficiency problem to be solved is dynamic. Note that a problem can be static in this sense even if it occurs over a long period of time—as long as the benefits and costs do not change from period to period.

4. For some very useful treatments of discounting in an environmental context, see Goulder and Stavins (2002), Brennan (1999), and Portney and Weyant (1999).

Chapter 3: The Benefits and Costs of Environmental Protection

1. This discussion is based on Nathaniel O. Keohane, Benjamin Van Roy, and Richard J. Zeckhauser, "The Optimal Management of Environmental Quality with Stock and Flow Controls." AEI-Brookings Joint Center for Regulatory Studies Working Paper 05-09, June 2005.

2. Robert Costanza, Ralph d'Arge, Rudolf de Groot, Stephen Farber, Monica Grasso, Bruce Hannon, Karin Limburg, Shahid Naeem, Robert V. O'Neill, Jose Paruelo, Robert G. Raskin, Paul Sutton, and Marjan van den Belt, "The Value of the World's Ecosystem Services and Natural Capital," *Nature* 387:253–260 (15 May 1997).

3. See Michael Toman, "Why Not to Calculate the Value of the World's Ecosystem Services and Natural Capital," *Ecological Economics* 25:57–60 (1998), 58. Curiously, Costanza et al. acknowledge this very point at the beginning of their article. They write, "It is trivial to ask what is the value of the atmosphere to humankind, or what is the value of rocks and soil infrastructure as support systems" (255). But as we explain in the text, they end up confusing marginal and total effects.

4. This discussion is based on Albert L. Nichols, "Lead in Gasoline," in Richard D. Morgenstern, ed., *Economic Analyses at EPA: Assessing Regulatory Impact* (Washington, D.C.: Resources for the Future, 1997), 49–86.

5. For an interesting discussion of benefit-cost analysis as applied to the environment, see "Cost-Benefit Analysis: An Ethical Critique," by Steven Kelman, with replies by James Delong, Robert Solow, Gerard Butters, John Calfee, and Pauline Ippolito. *AEI Journal on Government and Social Regulation* (January/February 1981): 33–40. Reprinted in Robert N. Stavins, ed., *Economics of the Environment: Selected Readings*, 5th ed. (New York: Norton, 2005), 260–275.

6. Kenneth J. Arrow et al., "Is There a Role for Benefit-Cost Analysis in Environmental, Health, and Safety Regulation?" *Science* 272:221–222 (12 April 1996). Reprinted in Robert N. Stavins, ed., *Economics of the Environment: Selected Readings*, 5th ed. (New York: Norton, 2005), 249–254.

7. See Solow's response to Kelman, cited in note 5 above.

8. See Dallas Burtraw, Alan J. Krupnick, Erin Mansur, David Austin, and Deirdre Farrell, "The Costs and Benefits of Reducing Air Pollutants Related to Acid Rain," *Contemporary Economic Policy* 16:379–400 (October 1998).

9. This discussion is based on Andrew Metrick and Martin L. Weitzman, "Patterns of Behavior in Endangered Species Preservation," *Land Economics* 72(1):1–16 (February 1996); and Metrick and Weitzman, "Conflict and Choices in Biodiversity Preservation," *Journal of Economic Perspectives* 12(3):21–34 (Summer 1998).

10. It is worth pointing out that strict Pareto efficiency is much less attractive as a general welfare criterion for evaluating overall outcomes, such as a particular equilibrium

2. The two-period nonrenewable natural resources problem we solve here is originally due to Griffin and Steele (1980) and is modeled on the interpretation in Teitenberg (2003, 89–93).
3. Notice that this problem is constructed so as to exhaust the stock of oil entirely. In reality, it may not be efficient to completely exhaust many nonrenewable resource stocks because the cost of extracting these resources can become very high as stocks dwindle. (Recall that we have assumed that marginal extraction costs were constant in the two-period problem.)
4. Among other conditions, the Hotelling Rule relies on the assumptions that marginal extraction costs and ore grades are constant. If these assumptions are not met, the Hotelling model is somewhat more complicated, although the general "no arbitrage" feel of the model is preserved.
5. See Allen L. Torell, James D. Libbin, and Michael D. Miller. "The Market Value of Water in the Ogallala Aquifer," *Land Economics* 66(2):163–175 (1990).

Chapter 7: Stocks that Grow
1. The following two-period problem is adapted from Barry C. Field, *Natural Resource Economics: An Introduction* (Long Grove, IL: Waveland Press, Inc., 2001).
2. Figure 7.4 is adapted from Field (2001), 232.
3. See Jonathan Rubin, Gloria Helfand, and John Loomis, "A Benefit-Cost Analysis of the Northern Spotted Owl," *Journal of Forestry* 89:25–30 (December 1991).
4. See D. Hagen, J. Vincent, and D. Welle, "Benefits of Preserving Old-Growth Forests and the Spotted Owl," *Contemporary Policy Issues* 10:13–26 (April 1992).
5. van Kooten, G. Cornelius, Clark S. Binkley, and Gregg Delcourt, "Effect of Carbon Taxes and Subsidies on Optimal Forest Rotation Age and Supply of Carbon Services," *American Journal of Agricultural Economics* 77(2):365–374 (1995).
6. See John Creedy and Anke D. Wurzbacher, "The Economic Value of a Forested Catchment with Timber, Water and Carbon Sequestration Benefits," *Ecological Economics* 38(1):71–83 (2001).
7. See Tom Tietenberg, *Environmental and Natural Resource Economics*, 6th ed., (Reading, MA: Addison Wesley Longman, Inc., 2003), 264.
8. See Robert Repetto and M. Gillis, *Public Policy and the Misuse of Forest Resources*, (Cambridge, UK: Cambridge University Press, 1988).
9. See Charles Wood and Robert Walker, "Saving the Trees by Helping the Poor: A Look at Small Producers along Brazil's Transamazon Highway," *Resources* 136 (Summer):14–17 (1999).
10. See Robert T. Deacon, "Deforestation and the Rule of Law in a Cross-Section of Countries," *Land Economics* 70:414–430 (1994).
11. See Robert L. Mendelsohn, "Property Rights and Tropical Deforestation," *Oxford Economic Papers* 46:750–756 (1994).
12. See Charles Wood and Robert Walker, "Saving the Trees by Helping the Poor: A Look at Small Producers along Brazil's Transamazon Highway," *Resources* 136:14–17 (Summer 1999).
13. The model is due originally to M. B. Schaefer, "Some Considerations of Population Dynamics and Economics in Relation to the Management of Marine Fisheries," *Journal of the Fisheries Research Board of Canada* 14:669–681 (1957).

14. The steady-state analysis, while a simplification, provides a good basis from which to develop the intuition and general conclusions of the bioeconomic fishing model. Nonetheless, some important aspects of the problem will be lost in our departure from the dynamic context. In particular, the simplification erases the link between today's harvest and tomorrow's stock, as well as the time value of money. We will be more explicit about the costs of the steady-state assumption later in the chapter.

15. This is simply an assumption that the yield-effort function is a multiplicative function of the stock and the level of effort. A common yield-effort function is the constant returns per unit effort (CPUE) function: $Y = qXE$, where Y is yield, X is the fish stock, E is the level of fishing effort, and $q > 0$ is a "catchability coefficient."

16. At the efficient level of fishing effort, the slopes of the total benefit and total cost curves are equal, so marginal benefit equals marginal cost. If this is unclear, see our discussion of the relationship between total and marginal cost and benefit curves in chapter 2.

17. For a fascinating discussion of common-property resources, based on a vast body of research on community-level natural resource management, see Elinor Ostrom, *Governing the Commons: The Evolution of Institutions for Collective Action* (New York: Cambridge University Press, 1990).

18. For an exposition of the dynamic fishing model, see chapter 3 in Jon M. Conrad, *Resource Economics*, (New York: Cambridge University Press, 1999).

19. The following is summarized from F. Berkes, D. Feeny, B. J. McCay, and J. M. Acheson, "The Benefits of the Commons," *Nature* 340(6229):91–93 (1989).

20. See Food and Agriculture Organization of the United Nations, "The State of World Fisheries and Aquaculture 2002," available at www.fao.org, accessed September 16, 2004.

21. For a wonderful description of the initial abundance of these stocks, and their subsequent decimation, see Mark Kurlansky, *Cod: A Biography of the Fish that Changed the World*, (New York: Penguin Books, 1997).

22. See Suzanne Iudicello, Michael Weber, and Robert Wieland, *Fish, Markets and Fishermen: The Economics of Overfishing*, (Washington, D.C.: Island Press, 1999).

23. Annual losses in net benefits in the Bering Sea fishery at the time were estimated at $124 million, see Daniel Huppert, "Managing the Groundfish Fisheries of Alaska: History and Prospects," *Reviews in Aquatic Sciences* 4(4):339–373 (1991).

24. See National Marine Fisheries Service, *Our Living Oceans: Economic Status of U.S. Fisheries 1996*, available at www.noaa.nmfs.gov, accessed September 20, 2004.

25. The following is summarized from John M. Ward and Jon G. Sutinen, "Vessel Entry-Exit Behavior in the Gulf of Mexico Shrimp Fishery," *American Journal of Agricultural Economics* 76:916–923 (1994).

26. See Food and Agriculture Organization of the United Nations, "Marine Fisheries and the Law of the Sea: A Decade of Change," special chapter (rev.) in *The State of Food and Agriculture 1992*, FAO Fisheries Circular No. 853, FAO, Rome, 1993.

27. See National Marine Fisheries Service, *Report to Congress: Status of Fisheries of the United States*, (Siver Spring, MD: NMFS, 1998).

28. See National Research Council, *Dolphins and the Tuna Industry* (Washington, D.C.: National Academy Press, 1992).

29. See Mario F. Teisl, Brian Roe, and Robert L. Hicks, "Can Eco-Labels Tune a Market? Evidence from Dolphin-Safe Labeling," *Journal of Environmental Economics and Management* 43:339–359 (2000).

Chapter 8: Principles of Market-Based Environmental Policy

1. Ronald Coase, "The Problem of Social Cost," *Journal of Law and Economics* 3:1–44 (1969).

2. Coase, of course, realized the importance of transactions costs in the real world. Indeed, much of his (very long) article is devoted not to the so-called Coase Theorem, but rather to the proposition that even when transactions costs are sizeable—so that assigning liability does matter—government action is not necessarily preferable to laissez faire. For example, Coase argued that taxing a polluting factory created its own problems: The resulting reduction in smoke would attract more firms and residents to the area, raising the subsequent damages from pollution and lowering the factory's productivity. Note how narrow this argument is: It implicitly treats a single factory (rather than the industry) as the object of government policy, and relies on the quaint assumption that smoke from a factory only affects a well-defined region in its vicinity. More importantly, the argument misunderstands the nature of the tax remedy. The proper tax is equal to marginal damages *at the efficient point*, rather than (say) to marginal damages at the unfettered level of production or under some other arbitrary condition.

3. This discussion is based on Danièle Perrot-Maître and Patsy Davis, "Case Studies of Markets and Innovative Financial Mechanisms for Water Services from Forests," mimeo produced by ForestTrends and the Katoomba Group, May 2001.

4. Of course, the firm would prefer to receive a subsidy rather than to pay a tax. Indeed, that points to a potential problem in the long run: A subsidy would create too much incentive for firms to enter the regulated market. In contrast, while a tax might encourage firms to exit the industry, this is not distortionary (at least if the tax is set according to marginal damages), since only the very dirtiest firms with the most expensive abatement options will choose to exit rather than to pay the tax—and those are the firms for which the damages from pollution are greater than the social benefits from production.

5. Two sources for further reading on information-provision programs and certification are James T. Hamilton, *Regulation through Revelation: The Origin, Politics, and Impacts of the Toxics Release Inventory Program* (New York: Cambridge University Press, 2005), on the TRI program; and Benjamin Cashore, Graeme Auld, and Deanna Newsom, *Governing through Markets: Forest Certification and the Emergence of Non-State Authority* (New Haven, CT: Yale University Press, 2004), on forest certification.

6. This discussion is based on research by Ian W. H. Parry and Kenneth A. Small, "Does Britain or the United States Have the Right Gasoline Tax?," Resources for the Future Discussion Paper 02-12 (March 2002), available at www.rff.org.

7. You may recall from our discussion in chapter 5 that some provision of public goods will typically arise even in the absence of government intervention. In our two-person example of the flower garden, the neighbor with the higher value for the garden ends up providing a positive amount of the good. But remember that in real-world cases of interest, where large populations share the public good (e.g., clean air), free riding will dominate, and private provision of the public good will be much smaller than the efficient level.

8. Information on the EU-ETS in this section comes from Joseph Kruger and William A. Pizer, "The EU Emissions Trading Directive: Opportunities and Potential Pitfalls,"

Resources for the Future Discussion Paper 04-24 (April 2004), 57 pp., available at http://www.rff.org; Karan Capoor and Philippe Ambrosi, "State and Trends of the Carbon Market 2006," (Washington, D.C.: World Bank, 2006), 49 pp.; and information from Point Carbon, available at www.pointcarbon.com.

9. Of course, privatizing the resource is not the only solution. As we saw in chapter 7, with the example of beaver hunting in James Bay, local communities can manage natural resources sustainably through social norms and customs of stewardship rather than private property. See also the work of Elinor Ostrom, cited in note 18 of chapter 7.

10. This discussion draws on James Y. K. Luk, "Electronic Road Pricing in Singapore," *Road & Transport Research* 8(4):28–40 (1999); the article "Electronic Road Pricing" in Wikipedia (http://en.wikipedia.org/wiki/Electronic_Road_Pricing), and information available from Singapore's Land Transport Authority at www.lta.gov.sg.

11. You may notice that the true marginal cost curves are simply parallel shifts of the expected (average) marginal cost curve. This might seem at first to be a very special case. As Weitzman showed rigorously, however, such parallel shifts are a fairly general way of modeling uncertainty—for reasons beyond the scope of this discussion.

12. Unfortunately, this is not necessarily the free lunch that it might seem to be. It turns out that environmental regulations may exacerbate the preexisting distortions caused by those other taxes. For example, a carbon tax (or a cap-and-trade program) would raise the price of gasoline. This effectively lowers real wages, by reducing purchasing power. The effect is much the same as an increase in the income tax, and therefore contributes to distortions in the labor market. Whether or not the efficiency gain from reducing other taxes offsets the hidden cost of the regulation (what economists call the "tax interaction effect") is an empirical question—that is, it can go one way or the other depending on the circumstances.

Chapter 9: The Case for Market-Based Instruments in the Real World

1. Although minimizing costs is not the same thing as maximizing net benefits, the two are closely related. If the benefits of pollution control depend only on the total amount of abatement (in other words, if the damages from pollution depend only on the total amount of pollution), and two policies achieve the same objective, the one with lower costs must be more efficient than the other (even if the goal itself is not fully efficient). Cost minimization is a necessary (though not sufficient) condition for efficiency. In other words, it is necessary to minimize costs in order to maximize benefits minus costs.

2. Two other things are worth noting about the condition for cost-effectiveness presented above. First, it is necessary but not sufficient. It is possible to imagine scenarios in which firms' marginal costs are equated, but total costs are not minimized—for example, if the capital costs of abatement were very high. Second, we have been careful to state that the equal-marginal-costs principle applies only to "all firms that abate pollution." It is possible that some firms with very high marginal abatement costs ought not to control pollution at all (at least from a cost-minimization standpoint).   ·

3. Recall that a supply curve in a product market only makes sense in a competitive industry, since it embodies the assumptions that individual producers are price-takers and that they seek to maximize profit. The same is true with an aggregate MAC curve: It only makes sense in a context in which all regulated firms take the price of abate-

ment (e.g., the tax) as given, and seek to minimize compliance costs. Indeed, the aggregate MAC curve essentially has cost-effectiveness built in: It can be thought of as tracing out the cost of abating pollution on the margin, when abatement is allocated cost-effectively among firms. In other words, the cost being measured corresponds to the *least-cost allocation of abatement* among firms. For this reason, it doesn't make much sense to think about the aggregate MAC curve under uniform standards, since (as we have already seen) uniform standards are not cost-effective in general.

4. In fact, we have seen this result already, in the Coase Theorem. Saying that the equilibrium allocation of tradeable allowances is unaffected by the initial allocation, as long as transactions costs are small, is just a restatement of Coase's insight that private bargaining in the absence of transactions costs will lead to efficient outcomes regardless of the assignment of property rights.

5. The cost of adoption—the capital cost of installing or building new control equipment—will remain implicit in our analysis. Note that we can safely assume that the cost of installing a particular technology does not depend on whether a firm faces a tax or a standard. Therefore, it will not affect our comparison among various policy instruments.

6. We have glossed over two subtle points which are worth mentioning in an aside. First, we have implicitly treated more technology as a good thing. In fact, from the perspective of economics, the question we should ask is not "Which policy leads to more investment in new technologies?" but rather "Which policy leads to more *efficient* investment in new technologies?" Answering this question turns out to be harder than it looks (which helps explain why we tackled the easier question in the text). Nonetheless, there are strong reasons to believe that market-based instruments will be superior to performance standards on efficiency grounds in the face of technological change. The real question is which market-based instrument—an emissions tax or a cap-and-trade policy—is preferable. The answer depends in large part on the relative slopes of marginal cost and benefit, as in the case of uncertainty over marginal cost discussed in chapter 8. Indeed, you might be able to see the intuitive connection between "cost uncertainty" and technological change. Second, we have assumed that the regulator does not anticipate the adoption of the new technology. You can think of this as corresponding to a scenario in which the form and stringency regulation remains set in stone over a reasonably long duration—long enough for new technologies to arise. This may not be a bad description of the real world. In the United States, for example, environmental laws are only infrequently revisited. The Clean Air Act has been amended only twice since 1970—once in 1977 and once in 1990. The Clean Water Act has seen virtually no significant changes since 1972.

7. Two nuances are worth noting. First, since we are assuming a fixed market price for fish, and have equated net benefits with net revenues, maximizing net benefits is equivalent to minimizing cost. We make the distinction here mainly for the purposes of intuition. Second, it might appear at first that an IFQ policy, by introducing quotas with a market price, would reduce the net benefits to fishers. In fact, however, that is true for a particular fisher only to the extent that she must buy more IFQs than she sells. Moreover, from the perspective of society as a whole, the value of IFQ transfers is a wash; what matters is the difference between the total value of the harvest (reflected in its market price) and the real costs of catching it.

8. The following example is due to R. Scott Farrow, Martin T. Schultz, Pinar Celikkol, and George L. Van Houtven, "Pollution Trading in Water Quality Limited Areas: Use of Benefits Assessment and Cost-Effective Trading Ratios," *Land Economics* 81(2):191–205 (2005).

9. For an interesting analysis of the NO_X trading program, including the problems stemming from its assumption of uniform mixing, see Meredith Fowlie, "Emissions Trading, Electricity Industry Restructuring, and Investment in Pollution Abatement," mimeo, University of Michigan (September 2006).

Chapter 10: Market-Based Instruments in Practice

1. If you are interested in reading more, some comprehensive studies of market-based policies in the real world include: Charles E. Kolstad and Jody Freeman, eds., *Moving to Markets in Environmental Regulation: Lessons from Twenty Years of Experience*, (New York: Oxford University Press, 2006); Robert N. Stavins, "Experience with Market-Based Environmental Policy Instruments," in Karl-Göran Mäler and Jeffrey R. Vincent, eds. *Handbook of Environmental Economics, Vol. 1* (Amsterdam: Elsevier Science B.V., 2003), 355–435; Thomas Sterner, *Policy Instruments for Environmental and Natural Resource Management*, (Resources for the Future: Washington D.C., 2003), 363, and Theodore Panayotou, *Instruments of Change: Motivating and Financing Sustainable Development* (London: Earthscan Publications, Ltd., 1998).

2. Congress anticipated this problem in the 1977 Clean Air Act Amendments, which declared that significant capital investments at existing plants, designed to extend their lifetimes, would now trigger the stringent federal regulation that applied to new sources. The so-called New Source Review process entailed a costly survey of planned investments that might trigger those regulatory standards. Enforcing this provision, however, has bedeviled subsequent administrations and stirred up fierce opposition from electric power plants. The line between routine maintenance and "significant" upgrades has been difficult to establish. As a result, electric utilities have been reluctant to make investments that would improve the operating efficiency of their plants. At the same time, the EPA has been largely unsuccessful in forcing utilities that made past investments to install new control equipment.

3. Many of the details about the allowance trading program created by the 1990 Clean Air Act are taken from the definitive overview provided by A. Denny Ellerman, Paul J. Joskow, Richard Schmalensee, Juan-Pablo Montero, and Elizabeth M. Bailey, *Markets for Clean Air: The U. S. Acid Rain Program* (New York: Cambridge U.P., 2000).

4. See Dallas Burtraw, Alan Krupnick, Erin Mansur, David Austin, and Deirdre Farrell, "Costs and Benefits of Reducing Air Pollutants Related to Acid Rain," *Contemporary Economic Policy* 16(4):379–400 (October 1998).

5. See Nathaniel O. Keohane, "Environmental Policy and the Choice of Abatement Technique: Evidence from Coal-Fired Power Plants," Yale University mimeo available at www.som.yale.edu/faculty/nok4/files/papers/scrubbers.pdf [on adoption decisions]; and David Popp, "Pollution Control Innovations and the Clean Air Act of 1990," *Journal of Policy Analysis and Management* 22(4):641–60 (Fall 2003; on patents and technological innovation).

6. In fact, a small fraction—roughly 3 percent—of each year's allowances have been auctioned off by the EPA. This provision was included in the trading program to overcome concerns that giving away all of the allowances would make it possible for the

incumbent firms to hoard their allowances, driving up the price for firms that wanted to enter. But in a sign of the political power of the regulated industry, the allowance revenue was credited to the electric utilities that had given up their allowances to be auctioned, rather than being retained by the government.

7. The geographical distribution of pollution under the allowance-trading program is discussed by Ellerman and his colleagues in chapter 5 of *Markets for Clean Air* (see note 5).

8. This argument has been made by Denny Ellerman, among others. See A. Denny Ellerman, "Are Cap-and-Trade Programs more Environmentally Effective than Conventional Regulation?" in Charles E. Kolstad and Jody Freeman, eds., *Moving to Markets in Environmental Regulation: Lessons from Twenty Years of Experience.* (New York: Oxford University Press, 2006), 48–62.

9. Recall from our discussion of fisheries in chapter 7 that fishing to MSY is economic— but not biological—overfishing; thus the TACs would have to be lower than their current levels to induce the efficient level of fishing effort.

10. At the program's inception, IFQs represented a right to an absolute amount of fish, rather than a percentage of the TAC. But this policy led to total shares that exceeded the TAC in several fisheries soon after the program was implemented. The government had to buy back almost sixteen metric tons of shares, at significant cost.

11. See John H. Annala, "New Zealand's ITQ System: Have the First Eight Years Been a Success or a Failure?" *Reviews in Fish Biology and Fisheries* 6:43–62 (1996).

12. The excluded thirty covered areas fished lightly for only a few species, not suspected to be at serious risk of overfishing. See John H. Annala, *Report from the Fishery Assessment Plenary, May 1994: Stock Assessments and Yield Estimates.* (Wellington, NZ: MAF Fisheries Greta Point, 1994), quoted in Annala (1996), 47.

13. All data on market performance, including quota and lease prices and market activity, is taken from Richard G. Newell, James N. Sanchirico, and Suzi Kerr, "Fishing quota markets," *Journal of Environmental Economics and Management* 49(3):437–462 (2005).

14. Note that this equity concern is distinct from a potential efficiency concern. If some market participants grow large enough that they can affect the market price of quotas, then efficiency will suffer—just as in any market where firms have market power. Recall our discussion of market failures at the end of chapter 5.

15. See James Sanchirico and Richard Newell, "Catching Market Efficiencies," *Resources* 150:8–11 (2003).

16. See Clement and Associates, *New Zealand Commercial Fisheries: The Guide to the Quota Management System.* (Tauranga, New Zealand, Clement & Associates, 1997); and P. Major, "Individual Transferable Quotas and Quota Management Systems: A Perspective from the New Zealand Experience," in *Limiting Access to Marine Fisheries: Keeping the Focus on Conservation*, ed. K. L. Gimbel, Center for Marine Conservation and World Wildlife Fund, Washington, D.C.: 98–106 , (1994).

17. These fees covered more than 80 percent of the program's cost in 1994–1995. See Suzanne Iudicello, Michael Weber, and Robert Wieland, *Fish, Markets and Fishermen: The Economics of Overfishing* (Island Press, Washington, D.C., 1999).

18. See Iudicello et al. (1999), 105.

19. See Christopher M. Dewees, "Effects of Individual Quota Systems on New Zealand and British Columbia Fisheries," *Ecological Applications* 8(1):S133–S138 (1998).

20. See R. W. Mayer, W. B. DeOreo, E. M. Opitz, J. C. Kiefer, W.Y. Davis, B. Dziegielewski,

and J. O. Nelson, *Residential End Uses of Water* (American Water Works Association Research Foundation, Denver, 1998).

21. See Mary E. Renwick and Richard D. Green, "Do Residential Water Demand Side Management Policies Measure Up? An Analysis of Eight California Water Agencies," *Journal of Environmental Economics* 40(1):37–55 (2000).

22. See Ellen M. Pint, "Household Responses to Increased Water Rates," *Land Economics* 75(2): 246–266 (1999).

23. More precisely, price elasticity measures the percentage change in demand that results from a 1 percent increase in price.

24. See Christopher Timmins, "Demand-Side Technology Standards under Inefficient Pricing Regimes: Are They Effective Water Conservation Tools in the Long Run?" *Environmental and Resource Economics* 26:107–124 (2003).

25. See Erin T. Mansur and Sheila M. Olmstead, "The Value of Scarce Water: Measuring the Inefficiency of Municipal Regulations," AEI-Brookings Joint Center Working Paper 06-01, January 2006.

26. For a comprehensive summary, see Hanna L. Breetz, Karen Fisher-Vanden, Laura Garzon, Hannah Jacobs, Kailin Kroetz, and Rebecca Terry, "Water Quality Trading and Offset Initiatives in the U.S.: A Comprehensive Survey," Database available at: www.dartmouth.edu/~kfv/waterqualitytradingdatabase.pdf (2004).

27. See Ross and Associates Environmental Consulting, Ltd. (2000), "Lower Boise River Effluent Trading Demonstration Project: Summary of Participant Recommendations for a Trading Framework," Prepared for the Idaho Division of Environmental Quality, September. Available at: www.deq.state.id.us/water/data_reports/surface_water/ tmdls/boise_river_lower/boise_river_lower_effluent_report.pdf.

28. See Robert Repetto, Roger C. Dower, Robin Jenkins, and Jacqueline Geoghegan, *Green Fees: How a Tax Shift Can Work for the Environment and the Economy* (World Resources Institute, Washington, D.C., 1992).

29. Sterner, Thomas, *Policy Instruments for Environmental and Natural Resource Management* (Resources for the Future, Washington, D.C., 2003), 363.

30. See Don Fullerton and Thomas C. Kinnaman, "Household Responses to Pricing Garbage by the Bag," *American Economic Review* 86(4):971–984 (1996).

31. See Robin R. Jenkins, Salvador A. Martinez, Karen Palmer, and Michael J. Podolsky, "The Determinants of Household Recycling: A Material-Specific Analysis of Recycling Program Features and Unit Pricing," *Journal of Environmental Economics and Management* 45:294–318 (2003).

32. The Brazilian example is summarized from Kenneth M. Chomitz, "Transferable Development Rights and Forest Protection: An Exploratory Analysis," *International Regional Science Review* 27(3):348–373, (2004).

33. See Virginia McConnell, Margaret Walls, and Elizabeth Kopits, "Zoning, Transferable Development Rights, and the Density of Development," RFF Discussion Paper 05-32, February, Resources for the Future, Washington, D.C., (2006).

34. Figures for 1993–2000 from National Research Council, *Compensating for Wetland Losses under the Clean Water Act* (National Academy Press, Washington, D.C., 2001). Figures on wetlands mitigation banks in 2001 and 2005 taken from Jessica Wilkinson and Jared Thompson, *2005 Status Report on Compensatory Mitigation in the United States* (Washington, D.C.: Environmental Law Institute, 2006), 110 pages. Credit prices from

Richard R. Rogoski, "Liquid Assets: New Breed of Bankers Deal with Wetlands," *Triangle Business Journal*, September 8, 2006.

35. The discussion of conservation banking draws on a report titled "Mitigation Banking as an Endangered Species Conservation Tool," a report by Environmental Defense (Washington, D.C.: 1999), 80 pp., downloaded from www.environmentaldefense.org/documents/146_mb.PDF#search=%22ESA%20endangered%20species%20banking%22.

Chapter 11: Sustainability and Economic Growth

1. See Joseph Stiglitz, "The Ethical Economist," a review of *The Moral Consequences of Economic Growth*, by Benjamin M. Friedman, *Foreign Affairs* (November/December): 128–134 (2005).

2. See Donella H. Meadows, D. L. Meadows, J. Randers, and W. W. Behrens, *The Limits to Growth* (New York: Universe Books, 1972); Donella H. Meadows, et al. 1992, *Beyond the Limits* (Post Mills, VT, Chelsea Green Publishing Company, 1992).

3. See Julian Simon, *The Ultimate Resource* (Princeton, NJ: Princeton University Press, 1981); Julian Simon, *The Ultimate Resource II* (Princeton, NJ: Princeton University Press, 1996).

4. See William D. Nordhaus, "Lethal Model 2: The Limits to Growth Revisited," *Brookings Papers on Economic Activity* (2):1–59 (1992).

5. See, for example, World Bank, *World Development Report 1992* (New York: Oxford University Press, 1992); Gene M. Grossman and Alan B. Krueger, "Economic Growth and the Environment," *Quarterly Journal of Economics* 110(2):353–377 (1995).

6. See William T. Harbaugh, Arik Levinson, and David Molloy Wilson, "Reexamining the Empirical Evidence for an Environmental Kuznets Curve," *Review of Economics and Statistics* 84(3):541–551 (2002).

7. See, for example, John List and Mitch Kunce, "Environmental Protection and Economic Growth: What Do the Residuals Tell Us?" *Land Economics* 76(2):267–282 (2000); Michael Greenstone, "Did the Clean Air Act Cause the Remarkable Decline in Sulfur Dioxide Concentrations?" *Journal of Environmental Economics and Management* 47(3):585–611 (2004).

8. See Nordhaus, "Lethal Model II" (1992); and Jeffrey A. Krautkraemer, "Economics of Scarcity: The State of the Debate," in David Simpson, Michael A. Toman, and Robert U. Ayres, eds., *Scarcity and Growth Revisited: Natural Resources and the Environment in the New Millennium*, (Washington, D.C.: RFF Press, 2005), 54–77.

9. For a summary of these trends, see Nordhaus, "Lethal Model II" (1992).

10. It is also possible that declining ore quality is contributing to some price decreases.

11. See Martin L. Weitzman, "Pricing the Limits to Growth from Minerals Depletion," *Quarterly Journal of Economics* 114(2):691–706 (1999).

12. See World Commission on Environment and Development, *Our Common Future*, Report of the United Nations World Commission on Environment and Development (Oxford: Oxford University Press, 1987).

13. Robert M. Solow, "Sustainability: An Economist's Perspective," J. Seward Johnson Lecture to the Marine Policy Center, 14 June, Woods Hole Oceanographic Institution, Woods Hole, Massachusetts. Reprinted in Robert N. Stavins, ed., *Economics of the Environment: Selected Readings*, 4th ed. (New York: W. W. Norton and Company, 1991), 131–138.

14. This idea is due to Robert N. Stavins, A. F. Wagner, and G. Wagner, "Interpreting Sustainability in Economic Terms: Dynamic Efficiency Plus Intergenerational Equity," *Economic Letters* 79:339–343 (2003).

15. Capital depreciation is simply the loss in value of an economy's capital equipment due to use and the passage of time. For example, this year's U.S. gross domestic product would include the total value of tractors produced by labor and property in the United States this year, net of the decrease in value of all tractors produced by labor and property in the United States in this and previous years.

16. See Arthur C. Pigou, *The Economics of Welfare* (London: Macmillan and Co., 1932), part 1, chapter 3, available at www.econlib.org/library/NPDBooks/Pigou/pgEW3.html.

17. Graham Davis, Interview for "Measuring our Worth," Living on Earth, National Public Radio (April 9, 2004).

18. See William Nordhaus and James Tobin, "Is Growth Obsolete," Cowles Foundation Working Paper 398, Yale University, New Haven, CT, reprinted from Milton Moss, ed., *The Measurement of Economic and Social Performance: Studies in Income and Wealth, Vol. 38*, National Bureau of Economic Research, Cambridge, MA (1973).

19. See Robert Repetto, William Magrath, Michael Wells, Christine Beer, and Fabrizio Rossini, *Wasting Assets: Natural Resources in the National Income Accounts*, (Washington, D.C.: World Resources Institute, 1989).

20. See William D. Nordhaus and Edward C. Kokkelenberg, eds. 1999. *Nature's Numbers: Expanding the National Economic Accounts to Include the Environment* (Washington, D.C.: National Academy Press, 1999).

21. See Martin L. Weitzman and Karl-Gustaf Löfgren, "On the Welfare Significance of Green Accounting as Taught by Parable," *Journal of Environmental Economics and Management* 32:139–153 (1992).

22 The following is summarized from Kenneth J. Arrow, Partha Dasgulpa, Lawrence Goulder, Gretchen Daily, Paul Erlich, Geoffrey Heal, Simon Levin, Karl-Göran Mäler, Stephen Schneider, David Starrett, and Brian Walker, "Are We Consuming Too Much?" *Journal of Economic Perspectives* 18(3): 147–172 (2004).

23. This phrase is due to Robert Solow, "An Almost Practical Step toward Sustainability," Lecture on the occasion of the 40th anniversary, Resources for the Future, Washington, D.C. (1992).

24. See A. B. Jaffe, S. R. Peterson, P. R. Portney, and R. N. Stavins, "Environmental Regulation and the Competitiveness of U.S. Manufacturing: What Does the Evidence Tell Us?" *Journal of Economic Literature* 33:132–63 (March 1995); and Arik Levinson, "Environmental Regulations and Industry Location: International and Domestic Evidence," in J. N. Bhagwati and R. E. Hudec, eds., *Fair Trade and Harmonization: Prerequisites for Free Trade? Vol. 1* (Cambridge, MA: MIT Press, 1996), 429–57.

25. See Smita B. Brunnermeier and Arik Levinson, "Examining the Evidence on Environmental Regulations and Industry Location," *Journal of Environment and Development* 13(1):6–41 (2004).

26. See Josh Ederington, Arik Levinson, and Jenny Minier, "Footloose and Pollution-Free," *Review of Economics and Statistics* 87(1):92–99 (2005).

27. See Arik Levinson, "Environmental Regulatory Competition: A Status Report and Some New Evidence," *National Tax Journal* 56(1):91–106 (2003).

28. See Jeffrey A. Frankel and Andrew K. Rose, "Is Trade Good or Bad for the Environment? Sorting Out the Causality," *Review of Economics and Statistics* 87(1):95–91 (2005).

29. For a succinct description of this point of view, see the first issue of the journal *Ecological Economics*, published in February 1989.

Chapter 12: Conclusion
1. "The invisible green hand," in "How many planets? A survey of the global environment," *The Economist* 6 July 2002:15–17.

Further Reading

Chapter 2

Brennan, Timothy J. 1999. "Discounting the Future: Economics and Ethics," in Wallace E. Oates, ed., *The RFF Reader in Environmental and Resource Management*, Resources for the Future, Washington, D.C., 35–41.

Goulder, Lawrence H. and Robert N. Stavins. 2002. "An Eye on the Future," *Nature* 419 (October):673–674.

Portney, Paul and John Weyant. 1999. *Discounting and Intergenerational Equity*, Resources for the Future, Washington, D.C.

Weitzman, Martin L. 1998. "Why the Far-Distant Future Should Be Discounted at Its Lowest Possible Rate," *Journal of Environmental Economics and Management* 36(3):201–208.

Chapter 3

Arrow, Kenneth J., Maureen L. Cropper, George C. Eads, Robert W. Hahn, Lester B. Lave, Roger G. Noll, Paul R. Portney, Milton Russell, Richard Schmalensee, V. Kerry Smith, and Robert N. Stavins. 1996. "Is There a Role for Benefit-Cost Analysis in Environmental, Health and Safety Regulation?" *Science* 272:221–222.

Burtraw, Dallas, Alan J. Krupnick, Erin Mansur, David Austin, and Deirdre Farrell. 1998. "The Costs and Benefits of Reducing Air Pollutants Related to Acid Rain," *Contemporary Economic Policy* 16:379–400.

Costanza, Robert, Ralph d'Arge, Rudolf de Groot, Stephen Farber, Monica Grasso, Bruce Hannon, Karin Limburg, Shahid Naeem, Robert V. O'Neill, Jose Paruelo, Robert G. Raskin, Paul Sutton, and Marjan van den Belt. 1997. "The Value of the World's Ecosystem Services and Natural Capital," *Nature* 387:253–260.

Freeman, A. Myrick III. 2003. *The Measurement of Environmental and Resource Values*, 2nd ed., Resources for the Future, Washington, D.C.

Kelman, Steven. 1981. "Cost-Benefit Analysis: An Ethical Critique," with replies by James Delong, Robert Solow, Gerard Butters, John Calfee, and Pauline Ippolito, *AEI Journal on Government and Social Regulation* (January/February):33–40. Reprinted in Robert N. Stavins, ed. (2005), *Economics of the Environment: Selected Readings*, 5th ed. (New York: Norton): 260–275.

Keohane, Nathaniel O., Benjamin Van Roy, and Richard J. Zeckhauser. 2005. "The Optimal Management of Environmental Quality with Stock and Flow Controls," AEI-Brookings Joint Center for Regulatory Studies Working Paper 05-09, June.

Metrick, Andrew and Martin L. Weitzman (1996), "Patterns of Behavior in Endangered Species Preservation," *Land Economics* 72(1): 1–16.

Metrick, Andrew and Martin L. Weitzman (1998), "Conflict and Choices in Biodiversity Preservation," *Journal of Economic Perspectives* 12(3): 21–34.

Nichols, Albert L. (1997), "Lead in Gasoline," in Richard D. Morgenstern, ed., *Economic Analyses at EPA: Assessing Regulatory Impact* (Washington, D.C.: Resources for the Future): 49–86.

Nordhaus, William D. (1999), "Discounting and Public Policies that Affect the Distant Future," in Paul R. Portney and John P. Weyant, eds., *Discounting and Intergenerational Equity*, Resources for the Future, Washington, D.C., 145–162.

Sen, Amartya K. 1970. *Collective Choice and Welfare*, San Francisco: Holden-Day.

Toman, Michael. 1998. "Why Not to Calculate the Value of the World's Ecosystem Services and Natural Capital," *Ecological Economics* 25:57–60.

Chapter 4

Mankiw, N. Gregory. 2007. *Principles of Microeconomics*, 4th ed., Thomson South-Western, Mason, OH.

Pindyck, Robert S. and Daniel L. Rubinfeld. 2004. *Microeconomics*, 6th ed., Pearson Prentice-Hall, Upper Saddle River, NJ.

Chapter 5

Baumol, William and Wallace Oates. 1988. *The Theory of Environmental Policy*, 2nd ed., New York: Cambridge University Press.

Hardin, Garrett. 1968. "The Tragedy of the Commons," *Science* 162(3859):1243–1248.

Gordon, H. Scott. 1954. "The Economic Theory of a Common Property Resource: The Fishery," *Journal of Political Economy* 62:124–142.

Stavins, Robert N., ed. 2005. *Economics of the Environment: Selected Readings*, 5th ed., Norton, New York.

Tietenberg, Tom H. 2003. *Environmental and Natural Resource Economics*, 6th ed., Addison-Wesley, Boston.

Chapter 6

Griffin, James M. and Henry B. Steele. 1980. *Energy Economics and Policy,* New York: Academic Press.

Heal, Geoffrey. 1998. *Valuing the Future: Economic Theory and Sustainability*, Columbia University Press, New York.

Hotelling, Harold. 1931. "The Economics of Exhaustible Resources," *Journal of Political Economy* 39(2):137–175.

Jevons, Stanley. 1866. *The Coal Question*, Macmillan, London.

Tietenberg, Tom. 2003. *Environmental and Natural Resource Economics*, 6th ed., Addison Wesley Longman, Inc., Reading, MA.

Torell, L. Allen, James D. Libbin, and Michael D. Miller. 1990. "The Market Value of Water in the Ogallala Aquifer," *Land Economics* 66(2):163–175.

Chapter 7

Berkes, F., D. Feeny, B. J. McCay, and J. M. Acheson. 1989. "The Benefits of the Commons," *Nature* 340(6229): 91–93.

Clark, Colin W. 1973. "Profit Maximization and the Extinction of Animal Species," *Journal of Political Economy* 81(4):950–961.

Conrad, Jon M. 1999. *Resource Economics*, Cambridge University Press, New York.

Creedy, John and Anke D. Wurzbacher. 2001. "The Economic Value of a Forested Catchment with Timber, Water and Carbon Sequestration Benefits," *Ecological Economics* 38(1):71–83.

Deacon, Robert T. 1994. "Deforestation and the Rule of Law in a Cross-Section of Countries," *Land Economics* 70:414–430.

Field, Barry. 2001. *Natural Resource Economics: An Introduction*, McGraw-Hill, New York.

Hagen, D., J. Vincent, and D. Welle. 1992. "Benefits of Preserving Old-Growth Forests and the Spotted Owl," *Contemporary Policy Issues* 10(April):13–26.

Iudicello, Suzanne, Michael Weber, and Robert Wieland. 1999. *Fish, Markets and Fishermen: The Economics of Overfishing*, Island Press, Washington, D.C.

Kurlansky, Mark. 1997. *Cod: A Biography of the Fish that Changed the World*, Penguin Books, New York.

Mendelsohn, Robert L. 1994. "Property Rights and Tropical Deforestation," *Oxford Economic Papers* 46:750–756.

Ostrom, Elinor. 1990. *Governing the Commons: The Evolution of Institutions for Collective Action*, Cambridge University Press, New York.

Repetto, Robert and M. Gillis. 1988. *Public Policy and the Misuse of Forest Resources*, Cambridge University Press, Cambridge, UK.

Rubin, Jonathan, Gloria Helfand, and John Loomis. 1991. "A Benefit-Cost Analysis of the Northern Spotted Owl," *Journal of Forestry* 89(December):25–30.

Schaefer, M. B. 1957. "Some Considerations of Population Dynamics and Economics in Relation to the Management of Marine Fisheries," *Journal of the Fisheries Research Board of Canada* 14:669–681.

Teisl, Mario F., Brian Roe, and Robert L. Hicks. 2000. "Can Eco-Labels Tune a Market? Evidence from Dolphin-Safe Labeling," *Journal of Environmental Economics and Management* 43:339–359.

van Kooten, G. Cornelius, Clark S. Binkley, and Gregg Delcourt. 1995. "Effect of Carbon Taxes and Subsidies on Optimal Forest Rotation Age and Supply of Carbon Services," *American Journal of Agricultural Economics* 77(2):365–374.

Tietenberg, Tom. 2003. *Environmental and Natural Resource Economics*, 6th ed., Addison Wesley Longman, Inc., Reading, MA.

Wood, Charles and Robert Walker. 1999. "Saving the Trees by Helping the Poor: A Look at Small Producers along Brazil's Transamazon Highway," *Resources* 136(Summer):14–17.

Chapter 8

Capoor, Karan and Philippe Ambrosi. 2006. "State and Trends of the Carbon Market 2006," World Bank, Washington, D.C.

Cashore, Benjamin, Graeme Auld, and Deanna Newsom. 2004. *Governing through Markets: Forest Certification and the Emergence of Non-State Authority*, Yale University Press, New Haven, CT.

Coase, Ronald. 1969. "The Problem of Social Cost," *Journal of Law and Economics* 3:1–44.

Dales, J. H. 1968. *Pollution, Property, and Prices: An Essay in Policy-Making and Economics,* University of Toronto Press, Toronto.

Goulder, Lawrence H. 1998. "Environmental Policy Making in a Second-Best Setting," *Journal of Applied Economics* 1(2):279–328.

Hahn, Robert W. 1989. "Economic Prescriptions for Environmental Problems: How the Patient Followed the Doctor's Orders," *Journal of Economic Perspectives* 3(2):95–114.

Hamilton, James T. 2005. *Regulation through Revelation: The Origin, Politics, and Impacts of the Toxics Release Inventory Program,* Cambridge University Press, New York.

Kruger, Joseph and William A. Pizer. 2004. "The EU Emissions Trading Directive: Opportunities and Potential Pitfalls," *Resources for the Future Discussion Paper* 04-24 (April), available at www.rff.org.

Parry, Ian W. H. and Kenneth A. Small. 2002. "Does Britain or the United States Have the Right Gasoline Tax?," *Resources for the Future Discussion Paper* 02-12 (March), available at www.rff.org.

Perrot-Maître, Danièle and Patsy Davis. 2001. "Case Studies of Markets and Innovative Financial Mechanisms for Water Services from Forests," mimeo produced by Forest-Trends and the Katoomba Group, May.

Pigou, Arthur C. 1920. *The Economics of Welfare,* Macmillan and Co., London.

Tietenberg, Tom H. 1990. "Economic Instruments for Environmental Regulation," *Oxford Review of Economic Policy* 6(1):17–33.

Weitzman, Martin L. 1973. "Prices v. Quantities," *Review of Economic Studies* 41:477–491.

Chapter 9

Fowlie, Meredith. 2006. "Emissions Trading, Electricity Industry Restructuring, and Investment in Pollution Abatement," mimeo, University of Michigan.

Farrow, R. Scott, Martin T. Schultz, Pinar Celikkol, and George L. Van Houtven. 2005. "Pollution Trading in Water Quality Limited Areas: Use of Benefits Assessment and Cost-Effective Trading Ratios," *Land Economics* 81(2):191–205.

Tietenberg, Tom H. 2006. *Emissions Trading: Principles and Practice,* 2nd ed., Resources for the Future, Washington, D.C.

Chapter 10

Jaffe, Adams B., Richard G. Newell, and Robert N. Stavins, 2003. "Technological Change and the Environment," in Karl-Göran Mäler and Jeffrey Vincent, eds., *Handbook of Environmental Economics,* Volume I, Elsevier Science, Amsterdam.

Stavins, Robert N. 2003. "Experience with Market-Based Environmental Policy Instruments," in Karl-Göran Mäler and Jeffrey Vincent, eds., *Handbook of Environmental Economics,* Volume I, Elsevier Science, Amsterdam.

Burtraw, Dallas, Alan Krupnick, Erin Mansur, David Austin, and Deirdre Farrell. 1998. "Costs and Benefits of Reducing Air Pollutants Related to Acid Rain," *Contemporary Economic Policy* 16(4):379–400.

Chomitz, Kenneth M. 2004. "Transferable Development Rights and Forest Protection: An Exploratory Analysis," *International Regional Science Review* 27(3):348–373.

Dewees, Christopher M. 1998. "Effects of Individual Quota Systems on New Zealand and British Columbia Fisheries," *Ecological Applications* 8(1):S133–S138.

Ellerman, A. Denny, Paul J. Joskow, Richard Schmalensee, Juan-Pablo Montero, and Elizabeth M. Bailey. 2000. *Markets for Clean Air: The U.S. Acid Rain Program*, Cambridge University Press, New York.

Fullerton, Don and Thomas C. Kinnaman. 1996. "Household Responses to Pricing Garbage by the Bag," *American Economic Review* 86(4):971–984.

Harrington, Winston, Richard D. Morgenstern, and Thomas Sterner, eds. 2004. *Choosing Environmental Policy: Comparing Instruments and Outcomes in the United States and Europe*, Resources for the Future, Washington, D.C.

Iudicello, Suzanne, Michael Weber, and Robert Wieland. 1999. *Fish, Markets and Fishermen: The Economics of Overfishing*, Island Press, Washington, D.C.

Jenkins, Robin R., Salvador A. Martinez, Karen Palmer, and Michael J. Podolsky. 2003. "The Determinants of Household Recycling: A Material-Specific Analysis of Recycling Program Features and Unit Pricing," *Journal of Environmental Economics and Management* 45:294–318.

Keohane, Nathaniel O. 2006. "Cost Savings from Allowance Trading in the 1990 Clean Air Act," in Charles E. Kolstad and Jody Freeman, eds., *Moving to Markets in Environmental Regulation: Lessons from Twenty Years of Experience*, Oxford University Press, New York.

Keohane, Nathaniel O. "Environmental Policy and the Choice of Abatement Technique: Evidence from Coal-Fired Power Plants," Yale University mimeo available at www.som.yale.edu/faculty/nok4/files/papers/scrubbers.pdf.

Popp, David. 2003. "Pollution Control Innovations and the Clean Air Act of 1990," *Journal of Policy Analysis and Management* 22(4):641–60.

Kolstad, Charles E. and Jody Freeman, eds. 2006, *Moving to Markets in Environmental Regulation: Lessons from Twenty Years of Experience*, Oxford University Press, New York.

Mansur, Erin T. and Sheila M. Olmstead. 2006. "The Value of Scarce Water: Measuring the Inefficiency of Municipal Regulations," AEI-Brookings Joint Center Working Paper 06-01, January.

McConnell, Virginia, Margaret Walls, and Elizabeth Kopits. 2006. "Zoning, Transferable Development Rights, and the Density of Development," RFF Discussion Paper 05-32, February, Resources for the Future, Washington, D.C.

Newell, Richard G., James N. Sanchirico, and Suzi Kerr. 2005. "Fishing Quota Markets," *Journal of Environmental Economics and Management* 49(3):437–462.

Panayotou, Theodore. 1998. *Instruments of Change: Motivating and Financing Sustainable Development*, Earthscan Publications, Ltd., London.

Pint, Ellen M. 1999. "Household Responses to Increased Water Rates," *Land Economics* 75(2):246–266.

Renwick, Mary E. and Richard D. Green. 2000. "Do Residential Water Demand Side Management Policies Measure Up? An Analysis of Eight California Water Agencies," *Journal of Environmental Economics* 40(1):37–55.

Repetto, Robert, Roger C. Dower, Robin Jenkins, and Jacqueline Geoghegan. 1992. *Green Fees: How a Tax Shift Can Work for the Environment and the Economy*, World Resources Institute, Washington, D.C.

Sanchirico, James and Richard Newell. 2003. "Catching Market Efficiencies," *Resources* 150:8–11.

Stavins, Robert N. 2003. "Experience with Market-Based Environmental Policy Instruments," in Karl-Göran Mäler and Jeffrey R. Vincent, eds. *Handbook of Environmental Economics*, Vol. 1, Elsevier Science B.V., Amsterdam, 355–435.

Sterner, Thomas. 2003. *Policy Instruments for Environmental and Natural Resource Management*, Resources for the Future, Washington, D.C., 363.

Christopher Timmins. 2003. "Demand-Side Technology Standards under Inefficient Pricing Regimes: Are They Effective Water Conservation Tools in the Long Run?" *Environmental and Resource Economics* 26:107–124.

Chapter 11

Arrow, Kenneth, Partha Dasgupta, Lawrence Goulder, Gretchen Daily, Paul Erlich, Geoffrey Heal, Simon Levin, Karl-Göran Mäler, Stephen Schneider, David Starrett, and Brian Walker. 2004. "Are We Consuming Too Much?" *Journal of Economic Perspectives* 18(3):147–172.

Brunnermeier, Smita B. and Arik Levinson. 2004. "Examining the Evidence on Environmental Regulations and Industry Location," *Journal of Environment and Development* 13(1):6–41.

Ederington, Josh, Arik Levinson, and Jenny Minier. 2005. "Footloose and Pollution-Free," *Review of Economics and Statistics* 87(1):92–99.

Frankel, Jeffrey A. and Andrew K. Rose. 2005. "Is Trade Good or Bad for the Environment? Sorting Out the Causality," *Review of Economics and Statistics* 87(1):95–91.

Friedman, Benjamin M. 2005. *The Moral Consequences of Economic Growth*, Knopf, New York.

Greenstone, Michael. 2004. "Did the Clean Air Act Cause the Remarkable Decline in Sulfur Dioxide Concentrations?" *Journal of Environmental Economics and Management* 47(3):585–611.

Grossman, Gene M. and Alan B. Krueger. 1995. "Economic Growth and the Environment," *Quarterly Journal of Economics* 110(2):353–377.

Harbaugh, William T., Arik Levinson, and David Molloy Wilson. 2002. "Reexamining the Empirical Evidence for an Environmental Kuznets Curve," *Review of Economics and Statistics* 84(3):541–551.

Jaffe, A. B., S. R. Peterson, P. R. Portney, and R. N. Stavins. 1995. "Environmental Regulation and the Competitiveness of U.S. Manufacturing: What Does the Evidence Tell Us?" *Journal of Economic Literature* 33(March):132–63.

Krautkraemer, Jeffrey A. 2005. "Economics of Scarcity: The State of the Debate," in David Simpson, Michael A. Toman, and Robert U. Ayres, eds., *Scarcity and Growth Revisited: Natural Resources and the Environment in the New Millennium*, RFF Press, Washington, D.C.), 54–77.

List, John and Mitch Kunce. 2000. "Environmental Protection and Economic Growth: What Do the Residuals Tell Us?" *Land Economics* 76(2):267–282.

Meadows, Donella H., D. L. Meadows, J. Randers, and W. W. Behrens. 1972. *The Limits to Growth*, Universe Books, New York.

Meadows, Donella H. et al. 1992. *Beyond the Limits*, Chelsea Green Publishing Company, Post Mills, VT.

Nordhaus, William D. 1992. "Lethal Model 2: The Limits to Growth Revisited," *Brookings Papers on Economic Activity* 1992(2):1–59.

Nordhaus, William D. and Edward C. Kokkelenberg, eds. 1999. *Nature's Numbers: Expanding the National Economic Accounts to Include the Environment*, National Academy Press, Washington, D.C.

Nordhaus, William and James Tobin. 1973. "Is Growth Obsolete," Cowles Foundation Working Paper 398, Yale University, New Haven, CT, reprinted from Milton Moss, ed., *The Measurement of Economic and Social Performance: Studies in Income and Wealth, Vol. 38*, National Bureau of Economic Research, Cambridge, MA.

Pigou, Arthur C. 1932. *The Economics of Welfare*, 4th ed., Macmillan and Co., London.

Repetto, Robert, William Magrath, Michael Wells, Christine Beer, and Fabrizio Rossini. 1989. *Wasting Assets: Natural Resources in the National Income Accounts*, World Resources Institute, Washington, D.C.

Schelling, Thomas C. 1997. "The Cost of Combating Global Warming: Facing the Trade-offs," *Foreign Affairs* 76(6):8–14.

Simon, Julian. 1981. *The Ultimate Resource*, Princeton University Press, Princeton, NJ.

Simon, Julian. 1996. *The Ultimate Resource II*, Princeton University Press, Princeton, NJ.

Slade, Margaret. 1987. "Natural Resources, Population Growth, and Economic Well-Being: Issues and Evidence," in D. Gale Johnson and Ronald D. Lee, eds., *Population Growth and Economic Development*, University of Wisconsin Press, Madison, WI, 331–372.

Solow, Robert. 1992. "An Almost Practical Step toward Sustainability," Lecture on the occasion of the 40th anniversary, Resources for the Future, Washington, D.C.

Solow, Robert M. 1991. "Sustainability: An Economist's Perspective," J. Seward Johnson Lecture to the Marine Policy Center, 14 June, Woods Hole Oceanographic Institution, Woods Hole, Massachusetts. Reprinted in Robert N. Stavins, ed., *Economics of the Environment: Selected Readings*, 4th ed., W. W. Norton and Company, New York, 131–138.

Stavins, Robert N., A. F. Wagner, and G. Wagner. 2003. "Interpreting Sustainability in Economic Terms: Dynamic Efficiency Plus Intergenerational Equity," *Economic Letters* 79:339–343.

Stiglitz, Joseph. 2005. "The Ethical Economist," a review of *The Moral Consequences of Economic Growth*, by Benjamin M. Friedman, *Foreign Affairs* (November/December):128–134.

Weitzman, Martin L. 1999. "Pricing the Limits to Growth from Minerals Depletion," *Quarterly Journal of Economics* 114(2):691–706.

Weitzman, Martin L. and Karl-Gustaf Löfgren. 1992. "On the Welfare Significance of Green Accounting as Taught by Parable," *Journal of Environmental Economics and Management* 32:139–153.

World Commission on Environment and Development. 1987. *Our Common Future*, Report of the United Nations World Commission on Environment and Development, Oxford University Press, Oxford.

Index